Praise for *Icebound*

'A fascinating modern telling of Barents's expeditions ... Pitzer presents a compelling narrative situated in the context of Dutch imperial ambition.' **Michael O'Donnell,** *Wall Street Journal*

'Andrea Pitzer's worthy and superb account keeps us enthralled to the last chilling word.' **Dean King, author of** *Skeletons of the Zahara*

'Engrossing ... Andrea Pitzer brings Barents' three harrowing expeditions to vivid life.' **Hampton Sides, author of** *In the Kingdom of Ice*

'A masterwork of narrative non-fiction.' **Mitchell Zuckoff, author of** *Frozen in Time*

'Narratives of frozen beards in polar hinterlands never lose their appeal. Most of the good stories have been told, but in *Icebound* Andrea Pitzer fills a gap, at least for the popular reader in English, with the story of the 16th-century Dutch mariner William Barents ... Pitzer's prose is robust, clear and sometimes elegant.' **Sara Wheeler,** *Spectator*

'Once Barents heads for the northern tip of Nova Zembla for the third time in May 1596, the story becomes dramatic, and dire... Still, the journey is filled with wonders: optical phenomena and mysteries unravelled, such as where European songbirds go each summer. Ms. Pitzer's descriptions of the region sing.' *Economist*

'For the 21st-century reader who's seen one too many photos of emaciated polar bears loping across melting permafrost, *Icebound* can read a little like paradise really, really lost ... Pitzer writes with care about the Arctic landscape Barents encountered ... *Icebound* is a reminder that there was once a time when things were unknown.' *New York Times*

'One of the most remarkable survival stories ... In Andrea Pitzer's telling, the utter relentlessness of the challenge to stay alive is never anything less than compelling.' **Peter Stark, author of** *Last Breath* **and** *Young Washington*

'Beyond thrilling. Beyond enthralling. I found this a tale so involving that I simply couldn't put it down.' **Martin W. Sandler, author of** *1919* **and** *The Impossible Rescue*

'Gives readers a new understanding of the phrase 'uncharted territory' ... Methodically researched and elegantly told.'
Beth Macy, author of *Dopesick* and *Factory Man*

'A true pleasure. In chapters that are vivid, fast-paced, and meticulously researched, Pitzer takes us into the fierce world of polar exploration before the age of Shackleton. Her account breathes life into the forgotten quest of William Barents to find a Northeast Passage.'
Michael F. Robinson, author of *The Coldest Crucible*

'Allows a glimpse into the true nature of human courage. This is a book you will not want to put down, except to catch your breath.'
William E. Glassley, author of *A Wilder Time*

'A gripping adventure tale that deserves an honoured place in the long bookshelf of volumes dealing with Arctic shipwrecks, winter ordeals, and survival struggles.' **David M. Shribman, *Boston Globe***

'*Icebound* deserves a place beside such classics as Alfred Lansing's *Endurance: Shackleton's Incredible Voyage* and Roland Huntford's *The Last Place on Earth: Scott and Amundsen – Their Race to the South Pole*' ***Booklist***

'Accomplishes for William Barents what the explorer could not do for himself: rescue his amazing life from the grip of the Arctic and centuries of hagiography. The Barents who appears in Pitzer's spyglass seems impressively close to the actual man: intensely bold, highly skilled, and catastrophically wrong.' **P. J. Capelotti, author of *The Greatest Show in the Arctic***

'Fascinating, bizarre, and very human ... A riveting account of lives drawn into a world that seems at once a dream and a nightmare.'
Blair Braverman, author of *Welcome to the Goddamn Ice Cube*

'A compelling story, full of danger, shipwrecks, and failure, but noble and heroic as well. Some in Barents's party didn't make it, but it was remarkable that so many managed to survive.'
Anthony Brandt, author of *The Man Who Ate His Boots*

ALSO BY ANDREA PITZER

The Secret History of Vladimir Nabokov

One Long Night: A Global History of Concentration Camps

Icebound

SHIPWRECKED AT THE
EDGE OF THE WORLD

ANDREA PITZER

SIMON &
SCHUSTER

London · New York · Sydney · Toronto · New Delhi

First published in the United States by Scriber, an imprint of Simon & Schuster Inc., 2021
First published in Great Britain by Simon & Schuster UK Ltd, 2021
This edition published in Great Britain by Simon & Schuster UK Ltd, 2021

1 3 5 7 9 10 8 6 4 2

Simon & Schuster UK Ltd
1st Floor
222 Gray's Inn Road
London WC1X 8HB

www.simonandschuster.co.uk
www.simonandschuster.com.au
www.simonandschuster.co.in

Simon & Schuster Australia, Sydney
Simon & Schuster India, New Delhi

A CIP catalogue record for this book is available from the British Library

Paperback ISBN: 978-1-4711-8276-1
eBook ISBN: 978-1-4711-8275-4

Printed in the UK by CPI Group (UK) Ltd, Croydon, CR0 4YY

Photo Credits: Chapter openings are illustrations from editions of Gerrit de Veer's journal.
Image in chapter 7 courtesy of the Rijksmuseum, Amsterdam. All other chapter
opening images courtesy of the National Maritime Museum, Amsterdam.

For Joe

Contents

List of Maps — xi

One: The Open Polar Sea — 1

Two: Off the Edge of the Map — 27

Three: Death in the Arctic — 65

Four: Sailing for the Pole — 95

Five: Castaways — 125

Six: The Safe House — 147

Seven: The King of Nova Zembla — 171

Eight: The Midnight Sun and the False Dawn — 179

Nine: Escape — 199

Ten: Staggering Homeward — 225

Coda: The Shores of Nova Zembla — 265

Acknowledgments — 275

Notes — 281

Index — 287

List of Maps

The Dutch Republic, 1590s   8

The First Voyage, 1594: Amsterdam to Kildin Island   30

The First Voyage Out, 1594: Kildin Island to Nova Zembla   34

The First Voyage Home, 1594: Nova Zembla to Kildin Island   46

The Second Voyage Out, 1595: The Return to Vaigach Strait   72

The Third Voyage Out, 1596: Discovering Spitsbergen   102

The Third Voyage Out, 1596: Amsterdam to Ice Harbor   111

The Third Voyage Out, 1596: The Path to Ice Harbor   116

The Third Voyage, 1597: Sailing for Home   212

The Third Voyage Home, 1597: Nova Zembla to
Kildin Island   242

The Third Voyage, 1597: Kildin Island to Kola   254

Icebound

CHAPTER ONE

The Open Polar Sea

In 1594, while Spain laid siege to the Netherlands in the third decade of a bloody war, Dutch navigator William Barents prepared to sail off the edge of the known world. He would leave in the spring for distant Nova Zembla, whose shores stretched hundreds of miles above the Russian mainland. He intended to follow its coastline north as far as he could go.

Money drove every part of the project, providing both the means and the goal of the expedition: investors were looking to discover a northern trade route to China. But the voyage might also answer fundamental questions about the Earth. Was Nova Zembla—meaning "New Land"—an island that could be circumnavigated, or was it part of a polar continent that would make a northeast passage impossible? The former might mean lucrative trade with the Far East. The latter would mean vast new lands to discover.

No ship in recorded history had ever sailed north over Nova Zembla, or that far north above Europe at all. As one half of a small fleet charged with mapping territory in waters known and unknown, William Barents intended to embrace this challenge and head into uncharted waters. Meanwhile, two other vessels would sail south of Nova Zembla, close to the mainland. The southern route had been tried by other explorers, but so far none had reached China.

Barents's home country, the Dutch Republic, was just over a decade old at that moment. Across the next century, it would become the world's leading economic and naval power. It would surpass all other countries in shipbuilding. The transcendent art of Rembrandt and Vermeer would appear. A flowering of the spice trade and slavery would also sustain the country through decade after decade of a forever war during the attempt to defeat Spain. The art, the war, the slaves, and the spices would all combine to transform the young, tiny republic into a country as powerful as any on the planet. William Barents would play a role in that drama, but as he readied himself for his first voyage into the Arctic, his country was a blank slate, its sins and achievements still unwritten.

The historical record on William Barents before this moment is nearly as blank. Born in the northern Netherlands near the midpoint of the sixteenth century, he was likely in his forties when he left home on his first Arctic voyage, but the year of his birth is unknown. He had trained as a steersman and had a childhood fascination with maps that endured into middle age. "I always had the inclination, from my youth onward," he wrote of himself, "to use all my qualities to portray in maps the lands that I roamed and sailed with all the surrounding seas and waters." His most famous portrait shows a receding hairline, dark hair, sloping shoulders, and a nose like a chisel. Given the image's creation three centuries after Barents's death, maritime historian Diederick Wildeman has suggested that any connection to the historical Barents is dubious.

With no aristocratic pedigree, it's apparent from his writing that Barents had nonetheless gotten an education. Before he ever made preparations for the Arctic, he'd likely sailed all the shores of Western Europe, from the Baltic Sea to the Portuguese coast. At a time when war led most Dutch sailors to turn away from the Strait of Gibraltar between Spain and Morocco, Barents pioneered mapmaking in the region. His *New Description and Atlas of the Mediterranean Sea*, written with the influential preacher and geographer Petrus Plancius, would soon make its way into print.[1] Yet as he walked the long canals of Amsterdam, his voyages to the Mediterranean all lay behind him. He would never unfurl a sail in Spanish waters or see the coast of Italy again.

If his country's future wasn't yet written, he was as good a candidate as any to invent it. He couldn't yet know—no one in the port, in the city, in the country that day could know—that after his death he'd be memorialized across the centuries. His name would appear everywhere: on the streets of countless cities and towns in his home country, as well as California, Bulgaria, and Bashkortostan—even on the door of a hotel bar in Longyearbyen, the northernmost town in the world.

But he wasn't famous yet. He was simply a navigator hired to carry out commercial exploration thought up by others and backed by men of wealth. As he stood on the docks about to begin his long, painful trek into immortality, he was so far from fame that history would not even record the name of his boat.

• • •

War between the Dutch Republic and Spain had begun in Barents's youth, and would last until the end of his life. Spain's claim to the Low Countries undergirded every part of the rebellion. In 1567, Spain sent the third duke of Alva to the Netherlands with an army of ten thousand men to establish order. That September, the count of Egmont—who'd previously fought in battle for Spain and been knighted—was arrested. Condemned for treason over his reluctance to punish his own people, he was beheaded the following June.

In the name of Roman Catholicism, Spanish troops laid waste to the port city of Antwerp and besieged Haarlem farther north before slaughtering some two thousand soldiers in a series of massacres that became known as the Spanish Fury. At the eastern city of Zutphen, Dutch rebels looted churches and killed priests. When Spain regained the town in the dead of winter, troops got their revenge, drowning some five hundred people by pushing them into the frozen river through holes in the ice.

In 1576, the atrocities drove all seventeen provinces of the Netherlands to briefly join together and oppose Spain. Though Dutch forces couldn't oust their rulers from the Low Countries, the Spanish crackdown likewise failed to rout the rebels. In 1581, the Dutch renounced loyalty to the king of Spain, stating in the Act of Abjuration that they rejected "being enslaved by the Spaniards" and would "pursue such methods as appear to us to most likely secure our ancient liberties and privileges."

An independent republic was declared the same year. William, the wealthy prince of Orange, who'd long functioned as the symbolic leader of the revolt, was unable to make more than fleeting gains on the ground. He would be assassinated three years later.

The rebels' bald assertion of human liberty in the face of tyrannical rule predated the American Declaration of Independence by nearly two hundred years and was no better received by the monarch at which it was directed. Antwerp became the de facto capital of the rebellious territory, but the Spanish laid siege to the city for more than a year, encircling it and blocking the river that led to its gates. Eventually, in 1585, the confederacy of provinces surrendered the city. Under the terms of surrender, Protestants had a generous four years to relocate. Half of Antwerp's population of seventy-six thousand fled sooner rather than later, most to Amsterdam.[2]

Tens of thousands of additional religious refugees elsewhere on both sides of the conflict left for other nations or moved to towns within the Low Countries that were more sympathetic to their beliefs. An influx of Sephardic Jews arrived from Portugal, seeking religious freedom away from Spanish rule, further influencing the character of the north. These shifts would crystallize cultural and religious differences, driving a wedge between regions and shaping national identity.

The Dutch rebellion was Europe's first modern revolution against a monarch—then the first to reject monarchy itself. Unsettling to neighbors and adjacent royalty, Dutch resistance would become a blueprint and an inspiration for uprisings around the globe for the next three centuries. City and provincial identity remained strong, but some million and a half Dutch residents had gained a foothold on national independence.[3]

Securing that independence took nearly a century and tore their homeland in half. All of William Barents's adult life unfolded inside upheaval or war. Due to events far beyond his control, when he sailed into the Arctic, he would sail as a herald of the new Dutch nation.

• • •

A fleet had been formed. The expedition had financial backing. The monthslong process to plan a route, find ships, gather a crew, and provision vessels was complete. Now the voyagers would have to rely on a very old idea that Amsterdam geographers had embraced: that the North Pole would be warm. Though men had died, and whole ships had vanished into the ice, mapmakers claimed that beyond the frozen waste that cracked rudders and crushed hulls each winter, the polar region hid waters that might be sailed, even an open sea.

Across millennia, the idea of a navigable sea had tantalized explorers, opening up a wealth of possibilities. The Greeks had described an island "beyond the point where the north wind blows," with such a mild climate its people harvested crops twice a year.[4] In 1527, a merchant had written to King Henry VIII to argue that "sayling Northwarde and passing the pole" would prove shorter than any known route to India. The first part of the voyage would be treacherous, the argument went, but with sturdy vessels, an expedition breaking through the mountains of ice north of Europe would discover a sanctuary at the top of the world.

In May 1594, as he packed possessions for his first voyage into the Arctic, William Barents knew he hadn't invented the concept of a temperate North, which had haunted mapmakers for two thousand years—he was only the latest navigator to adopt it. But as one of only a few who would ever have a chance to try to prove it correct, he embraced the idea, pondering northern sea routes. Other earlier expeditions had set out for high latitudes, following the coastline along Norway to Russia, establishing relations with locals, or bringing promising news of new potential paths into the unknown. As of yet, none had found a northern passage to the far side of the world. Barents aimed to be the first.

The port of Amsterdam nestled behind a peninsula at the

southern end of the shallow bay known as the Zuiderzee. Outside the entrance to the Zuiderzee, a chain of islands ran parallel to the coastline, shielding the bay from the North Sea. Inside the bay, countless towns and landings carried on the brisk business of a rising nation. Shielded by fortifications and surrounded by a moat, Amsterdam sat the most protected among them.

Amid the bristling bustle of a harbor in wartime, a ship with three masts floated in the harbor. Stretching less than a hundred feet from bow to stern, it was nonetheless big enough to carry cargo. On this trip, however, hauling cargo wouldn't be the ship's central mission. And Barents's craft wouldn't be the largest setting out. But the master navigator *would* hold sway over his own journey and direct the ship's exploration into unmapped worlds. Embracing the idea of an open passage through the high Arctic to far Eastern empires, William Barents stood ready to risk vessels, crew, and his own life to prove its existence.

Barents set out in a time of cataclysmic change, with upheaval reshaping every corner of Dutch life. Violence and enterprise were transforming national identity, religion, government, industry, science, and art—all at once. No one thing could be separated from the other; wild revolution had worked its way down to the smallest details of existence. In the world that was emerging, each element was in flux.

Barents had begun exploration just as the Dutch dominated European shipbuilding. Though the craft was evolving, ships remained in that moment artisanal projects, in which each vessel was made by hand with little in the way of diagrams or written plans. Builders began with a set of blocks in a line on which they set the keel—the spine of the ship. Perpendicular to the keel, arcing planks known as ribs rose to breathe a shape into the cage of the hull. With the ribs in place, planks running parallel to the waterline could be attached,

NORTH
SEA

Texel

Enkhuizen

ZUIDERZEE

Amsterdam

Zeeland

N

The Dutch Republic, 1590s

and L-shaped knees set inside to brace and bind the structure. Planks, keels, and ribs were all still cut and shaped by hand. They had to be hammered and plugged, with joining pegs pounded in then cut flush to the exterior planks. One or more decks could be laid to divide the ship into levels, from the cargo hold at the very bottom of the ship; to the orlop in the middle, which held the guns and sleeping sailors; and the upper deck, which sat open to the elements topside. The "ceiling" of the ship—not the roof but the planks along the sides of the vessel—would finish off the interior.

The Dutch had just perfected the *fluyt*, a pear-bottomed craft meant for trade, not war. Stripped of armaments and the marines needed to man them, *fluyts* could carry twice the cargo of traditional merchant ships and be built at half the cost of other vessels. The lack of written plans meant design improvements couldn't be quickly cribbed and implemented by other countries, putting the Dutch at tremendous advantage for a time.

But beyond design improvements, the Dutch made their biggest advance by decreasing the time needed for preparing wood to use in shipbuilding. The year Barents set sail, the first patented windmill-driven saw went into service. A floating mill used wind power to run a large saw blade back and forth, delivering both greater precision and exponentially improved productivity. In a little over a decade, the entirety of the northern terrain across from Amsterdam harbor would be transformed into a vast industrial basin, with twenty shipyards in operation and more on the way.[5]

Barents's expedition couldn't command the most innovative or newest craft, but a state-of-the-art design was hardly necessary. Instead, he prepared to sail in a slightly older vessel of middling size—one maximized not for cargo but with portholes and cannons for use in potential battle with enemy ships.

Yet a shift was already underway. The first *fluyt* devoted to cargo instead of battle would be launched that year.[6] With the flexibility, speed, and economy of its shipbuilding ascendant, the new Dutch Republic found itself in the perfect position to stake a claim as a maritime empire.

Crews had readied three other ships for departure. With the republic a loose alliance of provinces still in the process of becoming a nation, individual cities and regions played a central role in backing new projects. The province of Zeeland, south of Amsterdam, would send the *Zwaan* (Swan). Enkhuizen, another city on the Zuiderzee less than thirty miles from Amsterdam, would provision the *Mercurius* (Mercury). And in addition to one full-size ship, Amsterdam would send a smaller vessel to help explore the coastline.

The question of the best course to sail had been disputed from the beginning. Barents and his mentor Petrus Plancius came down on the side of a high northern route, staking their reputations on the high Arctic being navigable.

The audacity of this idea was matched by the ambition of their upstart country. Any reliable sea route to the East would carry a flood of goods and money into North Sea ports, allowing the Dutch to establish a global presence and compete with existing European powers. Explorers from the Netherlands dreamed of sailing from Amsterdam north over Scandinavia and then Asia. Cresting the top of the world, they hoped to arrive at the kingdoms of China and Cathay. The latter had been mentioned in Marco Polo's travelogue centuries before and was thought to lie north of China proper. Europeans knew so little about the region that several more years would pass before they came to realize the two kingdoms were one and the same.

Meanwhile, the sea held dangers for everyone. A territory with hundreds of miles of coastline, the Netherlands in 1594 had yet to formally establish a national navy. Early in the war with Spain, a group of disenfranchised local noblemen and pirates called the Sea Beggars had harassed ships and raided Spanish vessels for goods. William of Orange had granted them letters of marque authorizing their piracy, and for a time they used England as a base for operations with the blessing of Queen Elizabeth I, a Protestant ruler who had no love for Spain.

Later, however, support for the Sea Beggars became too problematic for her, and in 1572 Elizabeth barred them from English shores. With no remaining port as a haven, they grew desperate. They sailed back home and managed to seize the Dutch town of Brielle from Spanish control, establishing the revolt as a force to be reckoned with. William of Orange's tactical accomplishments on the ground were few and far between, but the Sea Beggars' successes laid the foundation for future Dutch naval forces.

For the time being, in place of a national navy, individual towns and provinces had admiralty boards to defend their shores. By 1574, an admiralty had been formed at Rotterdam, but it was initially incapable of reliably protecting even merchant ships from pillage. Over the next fifteen years, other regional boards would be established, but coordination was poor and at times incoherent. For a time after the assassination of William of Orange, the leader of the Dutch forces was English and not Dutch.[7] Any convoy at sea had to be prepared to defend itself.

At that point, war—particularly endless, ramshackle conflicts like the Dutch revolt—tended to be bad for empires, bleeding them of military strength, destroying property, and emptying coffers. In fact, some violence unleashed in the Spanish Fury had been part of a mutiny sparked by imperial failure to pay troops garrisoned in the Netherlands. The Dutch, however, managed to create a new prototype, an early model of a military-industrial complex that allowed key sectors to thrive in wartime.

William Barents would benefit from the chaos that birthed the Netherlands. The population shift to escape religious violence combined with the flight of those rejecting monarchy to funnel wealth and entrepreneurs into northern provinces, triggering a cultural and financial boom. These provinces further built strong trade with Baltic states, exporting tile and bricks, and importing massive stores of grain to sell at home and abroad.

Concentrated capital, skilled tradesmen, and intellectuals gathered in the new republic. By the time Barents had returned from his Mediterranean voyages and prepared to set sail for the Arctic, the nation was filled with investors and merchants eager to find new markets for their goods. It was a historic convergence that brought the possibility of empire within the nation's grasp.

Yet Barents and his companions weren't the first to sail north. The idea of sailing into the high Arctic had roots reaching at least as far back as the ancient Greeks. In the fourth century BCE, the astronomer Pytheas wrote *On the Ocean*, an account of his unprecedented voyage high into the unknown.

Heading out from his home city of Massalia—later known as Marseille—Pytheas recounted his passage through the Mediterranean and what is today the Strait of Gibraltar out into the Atlantic Ocean. No Greek had ever sailed so far north. Pytheas went up the western coast of Europe and over to Britain, where he hiked on foot and described circumnavigating the islands. He made his way to the Orkney archipelago and sailed six more days to a far northern land he called Thule—possibly Iceland.

Pytheas embraced his project not only as a guide for navigation, but also as a way to understand and interpret the wider world he discovered. He wrote of the midnight sun and how the moon influenced the tides, and he declared that one day's sail north of Thule lay the "Congealed Sea." In this sea, he wrote, "neither earth, water, nor air exist separately, but a sort of concretion of all these, resembling a sea-lung in which the earth, the sea, and all things are suspended."[8]

In the centuries that followed, Greek geographers and Roman writers would catalogue Scandinavian coastal tribes, from the Finnei in the far north to the Geats and Swedes clustered in the south. But the Arctic by and large remained a place of mystery.

The people eventually known as the Sami had inhabited Scandinavian Arctic regions for thousands of years. And before that, another indigenous Siberian people had ventured from Asia hundreds of miles north of the continent over frozen seas, building homes and hunting polar bears on distant Zhokhov Island eight thousand years ago.[9]

But in the annals of European seafaring, another group would conquer history's imagination as the mythic explorers of the North: the Vikings. More than a thousand years after Pytheas sailed from Massalia, the North's wildest sailors staked their claim. Sweeping outward from Scandinavia, they explored, plundered, and colonized. From AD 780 to 1070, they set out for northern Africa, the Iberian Peninsula, and Britain, also voyaging eastward as far as what is today Ukraine.

The shallow draft on Viking ships meant that some vessels could sail in less than two feet of water. Warriors regularly surprised enemies by winding their way up rivers large and small, including some that couldn't be navigated by enemy vessels. In 845, Danish Vikings sailed one hundred twenty ships into the Seine and took Paris. No European soil seemed safe from them.

But the most dramatic Viking legends came from their far northern voyages, which added a shudder of violence and geographic isolation to exploratory sailing. On oceangoing ships, or *knarr*, the Vikings navigated their way to Iceland and Greenland, where they made even more dramatic history. Late in the tenth century, Thorvald Asvaldsson was exiled from Norway to the Hornstrandir settlement on Iceland for what was depicted loosely as "some killings." From there, his son Erik the Red found himself banished from Iceland after murdering his neighbor. Erik sailed westward to Greenland, where he named several features of the landscape after himself and led the settlement of two colonies. Legends recount how his son Leif Eriksson captained his way to the North Atlantic reaches of America, where remains of Viking settlements would later be found dating from 1000.

The Vikings made their own marine innovations, perfecting clinker-style shipbuilding. Splitting oak logs radially into wedges, they fastened the planks with iron rivets, each plank overlapping the last. The flexibility of the planks allowed a ship to ride rough seas without shattering the hull. But the inevitable gapping from planks as they rode the waves meant constant bailing in bad weather. On one trip to Greenland, Erik the Red arrived with only fourteen of the thirty-five ships he had when he set out.[10]

A sound ship was important, but navigation was just as critical. A vessel that lost its way in open water might never return home. From the earliest days of seafaring, sailors developed ways to ensure safe passage and to orient themselves on the face of the deep. The Vikings had a few strategies to find their way, but little theory to guide them. Yet these strategies allowed them to sail vast distances.

From their earliest voyages, they kept near land and used coasts and visible landmarks to navigate, employing one of the most enduring tools known to sailors: a plumb line, weighted to measure the water's depth and warn of shallow seafloors. They shared directions with fellow navigators based on time or visual cues, including directions from Hernam in Norway to southern Greenland that are indecipherable as a map and can only be understood from the vantage point of a sailor at sea: "Steer due West for Hvarf in Greenland. You will then pass Hjaltland so close that you may just see them in clear weather, and so close to the Faroe Islands that half of the mountain is under the water, and so close to Iceland that you may have birds and whale from there."[11]

On the ocean, landmarks could be hard to come by, and directions might emphasize the number of days to sail before reaching the next island, an unreliable way to measure distance, since wind and weather varied. Sometimes other tactics were necessary. In the *Landnámabók* epic, Flóki Vilgerdarson made his way to Iceland in the ninth century carrying three ravens on his ship. After the first was set free, it turned and flew back the way they'd come. When he released the second, it flew up into the air but later returned to the ship. Once the third bird was loosed, it headed forward from the ship in the direction they were sailing, and they followed it to land.

The Vikings likely had instruments of navigation as well. Late in their era of expansion, compasses might have come to them from China through overland trade. They could have carved round bearing dials to help them stay on course. They might have used a gnomon, a crudely constructed sundial that could establish rough latitude. Historical records also include tales of sunstones, which were said to show the location of the sun even on a cloudy or snowy day. But if the Vikings had these tools, none have survived intact. Only fragments of relics remain.

Whatever their navigation instruments, the Vikings knew the position of the North Star and how high it should ride in the sky as they drew close to home. The Viking ability to hop islands hundreds of miles in a journey, first from continental Europe to the Faroe Islands, then on to Iceland, Greenland, and North America—

creating settlements as they went—remains a staggering feat of exploration.

By the time Barents set out, more than five hundred years later, the Viking legacy lived on in Scandinavia and other parts of Europe. But no Viking cultural inheritance made William Barents's voyage possible at the dawn of the Dutch empire. Instead, he would find his way using knowledge adapted from Pytheas and the Greeks and a host of other civilizations going back thousands of years.

Barents's chief advantage over the Vikings lay in a clear understanding of latitude and a way to measure it. The Vikings likely had no idea of the roundness of the Earth or the demarcations of the planet, but these ideas were bound up in a formal logic and geometry that was second nature to Pytheas as he sailed into the Arctic. If the rough ball of the planet spins on an imaginary axis running through the North and South Poles, the equator lays horizontally at zero degrees like a belt across its middle. With a full circle measuring 360 degrees, a ship heading from the equator to the North Pole covers a 90-degree arc as it sails. The position between the equator and the Pole in degrees is a ship's latitude.

Knowing where a ship's harbor of departure sits on that arc of latitude and the location of the boat at any given moment make up more than half the art of being able to find a way home. If sailors left Amsterdam in 1594 knowing that the port sits 52 degrees north of the equator, and they were able to find their current latitude anywhere at sea, they could easily sail north or south until they reached the latitude of Amsterdam again. Once a boat reached the right latitude, even without a compass, the direction of the morning sun would show the sailors east and west, and for any European in the Atlantic, home would lie east toward the morning sun along that imaginary line of latitude. A sailor ready to return home would never be permanently lost.

The Greeks later determined that the farthest places from the equator where the sun is directly overhead at some point during the year sit at predictable distances north and south of the equator. As a result, along with the equator circling the Earth, they had added one line above and one below, both parallel to it. The northernmost extreme of the sun's travels was christened the Tropic of Cancer, and the southernmost band the Tropic of Capricorn.

And the ancients realized that because of the changes in the sun's position, there should be another line of latitude closer to each pole, beyond which it would be possible to see the sun at midnight during the summer, and for sunlight to vanish entirely during part of the winter. The Greeks named the Arctic Circle for the polar constellation that should always be visible inside it—Ursa Minor, or Little Bear. The "Arctic" in Arctic Circle comes from *arktikos kyklos*, or "circle of the bear"—not creatures on the ground but the stars in the sky.

By 1080 a group of Arab astronomers compiled the Tables of Toledo, using trigonometry to establish the sun's angle in the sky at noon each day of the year in that city. Four hundred years later in Lisbon, astronomer and rabbi Abraham Zacuto would calculate declinations for the Sun, the Moon, and five planets, finally making tables simple enough for mariners to carry and use aboard a ship.

Zacuto's tables would be one of the first documents created in Portugal on a printing press. Similar charts were recalculated or openly copied, finding translation and adoption right up to the Dutch edition published in 1580 that William Barents carried with him to sea.[12]

Arab astronomers worked to perfect astronomical instruments as well, such as the astrolabe, a stationary tool that could track heavenly bodies with dials mounted over a plate. By the end of the fifteenth century, a simpler mariner's astrolabe would be adapted for use at sea: a flat brass ring hung vertically with a rod attached by a pin at its center like a clock, with two small holes in the outside end of the rod through which sunlight would line up when the dial was adjusted to the height of the sun. The mariner's astrolabe allowed navigators to track the sun's height in combination with declination tables to sail with much greater accuracy, even without any knowledge of trigonometry. An astrolabe would accompany William Barents into the Arctic.

These were the forebears whose work made Barents's voyage possible. But their insights came as a result of geographical placement as much as the discoveries by individuals themselves. At the nexus of trading routes, Greeks and Muslims alike had benefited from collecting the most advanced knowledge coming from other cultures and places. During the five thousand years before Barents's birth, the Babylonians, Egyptians, and Indians, along with Muslim scientists from Persia to Spain, and Chinese inventors farther east all contributed to an evolving body of knowledge through trade and scientific advancements that would unlock the mysteries of heaven and earth for explorers.

But Europe had lived for most of a millennium largely absent the astronomical knowledge of the Greeks, and their reintegration of lost math and theory sometimes left them unprepared for the navigation challenges they embraced. In 1492, Christopher Columbus made the first of his voyages to the New World under Spanish auspices, finding not the Indies he expected but an entire Western Hemisphere that came as a complete surprise to him.

It seems astounding that he could make that mistake, but Greek mathematics and astronomy were garbled as they were carried forward and reinterpreted, with the result being that the globe of Christopher Columbus's imagination was one-third smaller than the real world.[13]

By the time Barents sailed north, the first circumnavigation of the Earth had been completed more than half a century before by Ferdinand Magellan and Juan Sebastián Elcano, who, like Columbus, sailed on behalf of Spain. Between them, the two navigators managed to cross the Atlantic, arrive at the Cape of the Eleven Thousand Virgins near the tip of South America, cross the Pacific to the Philippines, and return to Spain by rounding southern Africa—a route around the world that veered from middle to southern latitudes. It remained an open question whether an intrepid voyager might likewise skirt the continents by heading north.

Meanwhile, the Scientific Revolution in Europe gathered steam. William Barents left Amsterdam for the Arctic at almost the midway point between the 1543 publication of Copernicus's *On the Revolutions of the Heavenly Spheres* and Galileo's 1632 defense of it, which led to the latter being tried as a heretic. During Barents's lifetime, Danish astronomer Tycho Brahe struggled to decipher which heavenly bodies might revolve around the Earth and which around the sun. The sailors put this still incomplete but growing understanding of the heavens to good use, wielding scientific instruments and cartography to track everything they encountered, transforming new seas and terrain into maps.

Among Protestant nations, a call to search for a northern route via the North Pole had been taken up first by England. In 1527, Robert Thorne, who'd embraced the idea of an open polar sea, managed to persuade Henry VIII to fund "two fair ships well manned and victualled, having in them divers cunning men to seek stranger regions." The first vessel was reportedly forced to turn back, and the second one vanished.

The riches of the Far East remained the prize that drove exploration in all directions. England moved to stake a seafaring claim by sending John Cabot across the Atlantic in 1497 to North America to seek a westerly route to China. A 1553 expedition from England using hired explorers attempted to sail over Europe to China but led instead to trade with Russia. An attempt by sailor and former privateer Martin Frobisher two decades later to steer west instead of east and find an Arctic passage over North America appeared promising initially, despite several crewmen being kidnapped by Inuit.

Gold ore that Frobisher carried back to England from what is today Nunavut, Canada, in the farthest northern reaches of North America, sparked the interest of investors eager to finance subsequent voyages. They sought a charter from Queen Elizabeth I for the Company of Cathay, with the goal of founding a North American colony to mine gold and serve as a transit point for future expeditions to the Far East.

Frobisher's second westward voyage to the same region in 1577 involved collecting more ore. Expedition members also kidnapped two Inuit adults and a child, and claimed territory in the name of Elizabeth, who'd given it the name Meta Incognita (The Unknown Shore). Frobisher's third voyage to the area in 1578 dropped off prospective colonists, who were dismayed to find it snowing in July. By the time Frobisher began to prepare for his return voyage that August, the colonists had decided to abandon the settlement and head home with him.

Experts soon determined that despite initial test results, the mineral unearthed wasn't gold at all but fairly worthless amphibolite and pyroxenite. Three voyages to North America had led to nothing but ruin. With no easy northwest passage in sight, no colony, and no gold ore, the Company of Cathay went bankrupt.

Shortly after Frobisher went in search of a Northwest Passage, Arthur Pet and Charles Jackman headed east. Sent by England's Muscovy Company to "discover Cathay," they spent the summer of 1580 following the coastline over Norway toward Asia, sailing shores that were already frequented by the Pomor, Russian sailors who'd begun exploring northward centuries before.

Pet and Jackman met with indigenous people and managed to get through the strait between the Russian mainland and an island called Vaigach. Sitting below Nova Zembla, which lay farther north, the strait at Vaigach Island seemed to offer a viable path toward China, but the explorers' progress eventually ground to a halt. A diary entry for the voyage from July 27, 1580, explains, "At one in the afternoon, master Pet and master Jackman did conferre together what was to be done considering that the winde was goode for us and we not able to passe for ice."[14] Despite their small ships that sat high in the water—one had a crew of only five men and a boy—Pet and Jackman ran both vessels aground during the effort to return home.

During the first weeks of his voyage, Barents's course would mirror Pet and Jackman's early route. With that in mind, Barents carried among his personal belongings a Dutch translation of their journals. Their voyage had taken them farther north than any Western European had gone before. But Barents would go farther.

In the wake of Pet and Jackman's failed pilgrimage, the Dutch trader Balthazar de Moucheron had sent another ship out on the same quest along the coast of Russia. The expedition managed to establish overland trade between the Netherlands and Russia, but de Moucheron's agent on the voyage, Olivier Brunel, died on the Pechora River still trying to make his way eastward.

As a result, sailing to China by sticking close to Russian shores became an even more speculative idea. After the birth of the Dutch Republic, de Moucheron requested permission from the new nation to launch another voyage.

The difficulties encountered by prior explorers sailing eastward led some to suspect that staying close to the mainland and sailing through Vaigach Strait wouldn't offer a reliable route to China. Geographer Petrus Plancius settled on a course that would ignore the strait altogether, leaving behind even Vaigach Island. Barents would head north of Europe, then make his way east, sailing over the Arctic land of Nova Zembla.

While it wasn't the dominant theory, Plancius wasn't alone in his belief in a polar route. He'd initially adopted the hypothesis from the world's most celebrated mapmaker, Gerardus Mercator of Flanders. Already famous for his atlases and spinning globes, Mercator had long pondered northern routes to China and believed that polar regions offered a more direct path to the Far East. Mercator would die in the year of William Barents's first Arctic voyage, his tomb inscribed to honor his brilliance at "showing the heavens from the inside and the Earth from the outside."

Over centuries, the idea of a warm North Pole had also infiltrated the minds of adventurous sailors and merchants, who began to dream of an easily navigable sea, one that might carry them over the top of the world and deliver them to profitable lands. It's a mark of both the draw of the unknown and the solidity of Plancius's reputation as a geographer that both Barents and Dutch merchants took the idea seriously, because it would turn out to be a lethal delusion.

The enthusiasm of investor Balthazar de Moucheron, however, was contagious. The province of Zeeland bought out his interest in the expedition and provisioned one ship, Enkhuizen agreed to send a second, and Plancius convinced Amsterdam to sponsor a third ship, along with a reconnaissance craft. In the end, the provincial and city councils decided that two routes would be explored, with the Amsterdam vessels following one course, and the ships from Enkhuizen and Zeeland remaining together to follow the other.

The Dutch merchants of Enkhuizen assigned a representative named Jan Huygen van Linschoten to look after their interests and goods on the trip. Van Linschoten was less known for any Arctic theory than for his time spent in warmer latitudes. Born in Haarlem, a city thirteen miles west of Amsterdam, van Linschoten had moved to Spain while still a boy to live with his brother. At the age of twenty, he secured a post as secretary to the archbishop of Goa, on the western side of India. Spending several years there, he copied or memorized details from a century's worth of political intrigue and secret maps—including coastal dangers, soundings, and navigation charts for how to sail from Europe south around Africa to eastern lands. At nearly the same moment that Barents's new atlas of the Mediterranean was ready for print, van Linschoten's cache of secrets about southern navigation routes would go off in the Netherlands like a bomb, igniting a Dutch-Portuguese rivalry.

With newly established navies, merchant ships, and tradable commodities in place during the last years of the sixteenth century, the principal remaining allies needed by explorers were businessmen and political leaders who could support the cost of expeditions. Polar navigation promised a foundation on which to build a nation, and van Linschoten had exposed closely held intelligence on which the Portuguese had kept a monopoly for decades. He not only laid out their navigation routes but also provided political analysis of the strengths and weaknesses of various alliances. He'd effectively written a how-to manual on how to reach the East, destabilize imperial colonies, and, perhaps, conquer the world.

After stealing secrets for the Dutch Republic, van Linschoten would now head into the Arctic on the same expedition as Barents, aboard a ship named *Mercury*, invoking not only the speed of the messenger for the gods but also his temperamental nature. Both men were committed to finding a route to the Far East, but they'd soon find themselves in opposition about how to get there—along with many other things.

In his journal, van Linschoten wrote that the *Mercury* would join the *Swan* to head north of Norway then eastward to the narrow Vaigach Strait above the Russian mainland. Meanwhile, Barents would take the unnamed Amsterdam boat and its smaller scouting craft on a course north of Vaigach Island up the western coast of Nova Zembla, to see how close to the North Pole it might lead him. If this mysterious land were, as Barents hoped, an island, there should be a polar sea above it.

To this end, Barents had been given control of his boat, with backing from Amsterdam merchants and the blessing of Prince Maurice, the leader of the Dutch Republic. He was no military commander. But on this voyage, Barents held broad responsibility for the route the ship would sail, as well as the success or failure of his part of the expedition. The fleet wasn't expected to make its way to China on this exploratory voyage, but was charged with charting reliable seas that a full expedition could travel, perhaps the following year.

The *Swan*, the *Mercury*, and the vessels from Amsterdam sailed out from the walls of their home cities. Items carried into the Arctic with Barents included ship's biscuit, barrels of meat and beer, crowbars, axes, hand drills and hacksaws, scrub brushes, muskets, musket balls, cooking pots, spears, halberds, gunpowder horns, swords, knives, low-cut leather shoes, hatchets, and dreams of discovery. The vessels held enough to feed their crews for eight months, though they planned to supplement their provisions along the way.[15]

As they set forth, they knew some things. They knew how to set the sails on a tall ship to catch the wind. They knew how to steer.

They knew how to work wood, and hunt, and trap. Barents could reckon latitude and knew the stars, and those he didn't know, he had charts for. Sailors understood that icebergs haunted the northern regions and could stretch for miles. Sometimes rising more than two hundred feet above the waterline, they were capable of dwarfing vessels and the tiny human presence guiding them.

Barents and his fellow crew members knew some things, but it wasn't enough. They possessed no scientific understanding of gravity, no telescopes, and no calculus. Though they could find their latitude, they couldn't yet determine longitude from aboard a ship. They were centuries away from deciphering the germ theory of disease. More than a hundred years would pass before humanity would discover that lightning was electricity. Decades remained before doctors would realize that blood circulates in the body, and that a cell is the unit of life. As he sailed into the Arctic, Barents would, in time, encounter wonders and terrors without understanding most of the forces at play in his universe.

He'd likely never heard the groan and crack of ice above the creak of the ship, the noise that carries across the water before its source can be seen. The crew had never seen a polar bear, and hadn't yet learned how white bears could move almost invisibly in a landscape of ice. They knew of scurvy, a disease that created sore afflictions among sailors, but they didn't yet know its cause. Its cure would be identified, forgotten, rediscovered, and doubted again for another three centuries.

As Barents prepared to sail, he left behind a wife and five children who were dependent on his fortunes. In his forties by now, he'd already seen more of the world than most humans. Taking into account the life expectancy for Northern European nobility in the era—a social class higher than his own—he likely had only a few more years left to live.

The crew boarded the ship in the crowded bustle of Amsterdam harbor, amid the single-masted longboats with oarsmen, the larger cargo-carrying hulks, and the tiny yachts and rowboats carrying sailors to and from land. Barents's vessel was one of many ships preparing to sail. Each had its own possibilities and was charged with its own mission. But with this voyage north, the Dutch were staking a claim to the future. In the history that followed, no sailor aboard any of the vessels there that day would eclipse the head-strong navigator bound for the Arctic. As William Barents's crew left home behind, along with the burghers of Amsterdam and the port itself, they began to make their way outside the annals of all recorded voyages. The ship sailed north in fear and wonder to pierce the veil of the unknown world.

CHAPTER TWO

Off the Edge of the Map

The dispute over which route to take had sparked ill will that would linger on the expedition. Jan van Linschoten was deeply skeptical of the plan for two ships to round Nova Zembla from the north. While he considered Barents an expert navigator, he thought Plancius had misled the burghers of Amsterdam with a wild-eyed polar theory. According to van Linschoten, Plancius insisted that the high northern route was "surely, most certainly, and without a doubt" correct. The geographer had given "a thousand kinds of questionable examples" in making those arguments while declaring that the route south around Nova Zembla—the route van Linschoten's ship would take—"was not there at all."[1]

The plan for a two-pronged expedition, however, was set in stone. Barents's ship and its smaller companion slipped out of the harbor at Amsterdam on May 29, 1594, taking almost a week to get to the Dutch island of Texel on the North Sea. At Texel, the Amsterdam contingent found the other two ships waiting for them. The four vessels would, in theory, set out together. With Enkhuizen represented by van Linschoten, the province of Zeeland sponsored Cornelis Nay—captain of the *Swan* and the commander of the modest fleet.

Along with his helmsman, Nay had sailed the Norwegian coast and spent time in Russian waters. His cousin, who had also lived there, came aboard as a Russian interpreter. Given the possibilities of human contact on the mainland below Nova Zembla, Nay also brought along a young Slav who'd been studying in the Netherlands.

The next day, June 5, an east wind rose at their backs. It was a favorable sign for departure, but the Amsterdam ships weren't ready on time. The *Mercury* and *Swan* left without them. Barents soon came trailing after, chasing the other ships up the coast.

They prepared to sail into the open sea and had planned to stay together as far as Kildin Island, a well-charted location off the Russian mainland just past the kingdom of Norway. From there, according to prior agreement, Barents would split off and sail his ship and the scouting vessel "above Nova Zembla, that is, underneath the Polus Arcticus"—the North Pole. He would answer the question of whether Nova Zembla was part of a land mass stretching all the way to the top of the world.

With the route as far as Kildin Island on the Norwegian-Russian border fully mapped, and a straightforward passage ahead of them for the first weeks at sea, the weather remained the only variable. Aboard the *Mercury*, Jan van Linschoten wrote that he'd been overcome by a "burning desire" to sail to the Arctic. He'd survived the dangerous Portuguese route rounding the Cape of Good Hope twice coming and going from Asia, only to be chased by English pirates and see the ship's cargo lost on the way home. If a northern route to China could be discovered, the dangers of that roundabout Cape passage could be avoided. Van Linschoten was convinced that any northern route would be six times shorter.[2]

As the fleet followed the North Sea toward the Pole, the sun spiraled overhead. The life of a sailor at sea is a wheel of routine, and each ship quickly established a daily rhythm. Crews were divided into staggered groups eating meals, sleeping belowdeck, and keeping watch in rotation, bringing on ballast to weight and balance the ship, furling and unfurling sails, dropping and weighing anchor, cleaning the ship, repairing lines, patching sails, and mending clothes.

Above decks, the sun might bake everything stiff and dry, but in the orlop—above the cargo hold but below the main deck— it would be hard to avoid the reek of unwashed bodies in unrelieved proximity. The many languages of water surrounded them day and night, slapping the sides of the ship at anchor, spraying out from a hull cutting through open sea, then thundering down from overhead.

Once at sea, the men were subject to the captain's authority in every matter and had little hope of subverting his will except through mutiny. He could order a range of punishments for infractions from confinement to execution, with little or no accountability until the ship returned to its home port, if then. Voyages that required hundreds of able seamen might mean that sailors who were collared into work would have few skills upon setting out.

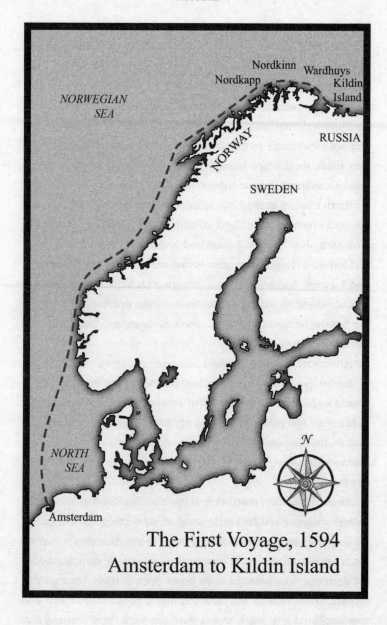

Nordkinn Wardhuys
Nordkapp Kildin
Island

NORWEGIAN SEA

RUSSIA

NORWAY

SWEDEN

NORTH SEA

N

Amsterdam

The First Voyage, 1594
Amsterdam to Kildin Island

But as seasoned sailors needing only small crews, the captains sailing with William Barents into the Arctic in 1594 had likely been able to visit local establishments in their home cities and secure sailors they knew or who'd been recommended to them.

The fleet headed out from Amsterdam with sixteen continuous hours of sunlight and outran more darkness every day until they sat far enough north that night was banished altogether. As they made their way up the western coast of Norway, the ships spied Trondheim and the islands of Lofoten, rounding the farthest northern reaches of the Norwegian coastline. They passed the jagged rocks leading up to the high plateaus at Nordkapp and Nordkinn, then the mounded cliffs of Stappane.

Despite occasional rough weather, they made good time and crossed from Norway to Russia. On the Russian side of the peninsula at Kildin Island, the fastest ships dropped anchor to wait for the rest of the fleet. By June 23, the ships were reunited. With the sun's circuit now entirely above the horizon, they could sail anytime the wind came to their aid.

Kildin had become a well-known trading station and a waypoint for sailors heading to Russia. The small outpost rose high out of the water on piles, flashing inhospitable steep faces and a rocky plateau that lay an hour's climb away from the accessible shores. But near its eastern end sat a peaceful harbor as protected as any city port. Lying just inside the Arctic Circle, Kildin had no trees or tall grasses, but occasional mosses and moor brambles sat low to the ground. There were said to be bears and wolves.

Ten miles long and some three miles across, Kildin was a crossroads for Europeans and Russians, as well as Finns and indigenous Sami people, who moved eastward onto the island during summer months. Sami clothes and shoes were made of reindeer fur, and reindeer the size of adult rams pulled the Sami sleds. Some of the

Dutchmen looked down on the Finns and Sami crowded into turf huts. Both groups made a living selling fish to the Russians, who dried it for resale to passing ships.

Disembarking from the *Mercury* to take stock of these groups, van Linschoten described them as dirty pigs, penniless, and misshapen, "living a poor miserable life." Though van Linschoten had lived in other lands among other peoples than his own, whatever curiosity or empathy he'd picked up wasn't extended to the Sami, whom he described as barely more than animals.

Despite van Linschoten's uncharitable bile, it wasn't the Sami who gave the Dutchmen trouble at Kildin. The captain of a Danish cargo ship came aboard the *Mercury*, wondering why the first ships of the fleet to arrive weren't continuing south into the White Sea above Russia, given the favorable winds. He demanded to see their papers, but they pretended to be confused over language issues, ignoring his request and sending him away.

They found themselves beset by Russians as well. A customs official for the grand duke of Moscow complained bitterly that they shouldn't fish in the waters at Kildin day after day without asking permission. His tone implied some kind of present was due. After visitors didn't offer a gift, the Russians took advantage of the midnight sun while the Dutchmen were asleep.

At one point in the bright night, the Dutch sailor on watch saw the Russians paddling furiously toward shore. In their small boat lay the fishing nets that Dutch sailors had cast against the customs officer's wishes. The thieves made it back to land, only to be chased by the watchman, who woke his shipmates and with them fell into the ship's small boat while yelling at the Russians. Landing just ahead of their pursuers, the Russians dropped their coats in the dinghy, the better to run on foot. The furious Dutchmen caught the fleeing thieves and beat them. Returning to shore, they rowed the Russian dingy back to their ship, bringing their nets and the Russians' coats with them.

The customs official was obliged to sail out to them the next day and plead with the captain for the return of the dinghy and coats. It wasn't possible to punish the Russians who'd stolen the nets, the officer explained, because they'd fled to the mountains. The Dutchmen relented and returned the Russians' belongings, and after that were left alone.

By Wednesday, June 29, the two groups had finished their last preparations. Events on Kildin had been unpredictable, but not too different in spirit from encounters that could have taken place in any port on any number of voyages. Now, however, the fleet prepared to split in two, with each group making its way into wilder terrain.

William Barents took the Amsterdam ship and its yacht and set a course northeast toward Nova Zembla. As he sailed away, Admiral Nay in the *Swan* and van Linschoten in the *Mercury* stayed together, heading nearly due east, aiming to stay below Nova Zembla and make for Vaigach Island. If all went well, Barents would round Nova Zembla from the north and Nay from the south. The two groups hoped to meet in a few weeks on the other side. But anything could happen in the coming days, and no one knew where wind and weather would take them.

Based on the instructions given to them by the States General, if they couldn't find each other on the eastern side of Nova Zembla, they vowed to set a course for Kildin and wait there for the rest of the fleet until the end of September, when they'd return home as a group. Given the ongoing conflict with Spain, it remained wiser to stay together as a show of force when sailing back toward Dutch waters.

The next day, Barents found a breeze to sail on, unfurling two

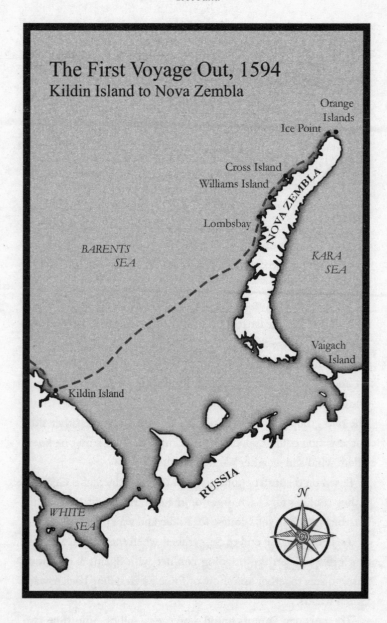

The First Voyage Out, 1594
Kildin Island to Nova Zembla

Orange
Islands
Ice Point

Cross Island
Williams Island

Lombsbay

NOVA ZEMBLA

BARENTS
SEA

KARA
SEA

Vaigach
Island

Kildin Island

RUSSIA

N

WHITE
SEA

large sails to catch it squarely. Tracking the seafloor under them as they sped along, the crew lowered a lead plumb line hardly more advanced than the one used by the Vikings. On June 30, the line sank more than five hundred feet into the depths without touching bottom. Every man aboard could have stood on the shoulders of the next sailor to make a tower, and still they wouldn't have covered even a third of the depth between the ship and the seafloor.

As they sailed east during their first week after leaving Kildin, they hauled in the plumb line to find it coated with black and white, gray, then red sand, along with broken shells. Their fourth day out, they were making good speed on a stiff wind from the southeast when a fog rose up around them. Trimming sails, they cast out the line again to be sure they wouldn't run aground while they were sailing half-blind. But once more, sailors found only a bottomless sea.

The North Sea near home had its difficult moments, but while sailing the waters between Kildin Island and Nova Zembla, the crew encountered new challenges. The relatively warm water sweeping in from the west on the last dregs of the gulf current crossing the Atlantic Ocean ran hard into colder water that lay to the east. Any given ship could find itself moving several directions at once on the choppy waves, and in high winds, sailing could become impossible.

On July 3, William Barents pulled out his cross-staff, a wooden pole with a sliding crossbar that could be used to determine the height of celestial bodies. Hoisting the main bar up between his eye and the horizon, he slid the crossbar along its length until, at great long-term risk to his retina, he looked into the sun and saw its lower rim touching the crossbar. After recording its height from the horizon, he reckoned his location and adjusted course.

The next day—not yet a week after setting out from Kildin—

they caught sight of land. William Barents had successfully crossed the sea that in time would bear his name. The western shore of Nova Zembla lay in view. They'd left Europe behind.

Barents shifted his course again and sailed toward unmapped terrain. The low coasts of the island, or islands—it was not yet clear whether Nova Zembla was an archipelago—sat strewn with earth and rock beaches some yards in from the shore, rising to flat plateaus above them, and sometimes tall hills or low mountains as well. Glaciers dotted the landscape, but in the short summer months, lichens or moss tinted the visible earth and sides of the cliffs with spring or autumn colors.

The Dutchmen came to a long spit of land they named Langenes and found a bay just to its east where they rowed a boat to land. But finding no people there, they returned to the ship and sailed on. They wound their way northeast up the coast, sailing into bays and inlets. Capo Baxo—their bastardization of Spanish for "Low Point"—and Lombsbay came next along the coast. At Lombsbay, they went ashore again, stepping into a bleak landscape, and found a coven of guillemots. As they waddled, the birds' charcoal heads and wings bobbed in sharp contrast with the white breasts puffed before them like shields.

Other signs of life—or former life—appeared as well. The scouting party came across an old mast lying on the ground and recognized the relic as part of a Russian Pomor ship. The spear of wrecked lumber could be taken as both good and bad news. Someone had sailed this far up the coast of Nova Zembla before, meaning the area was navigable. But the mast—or perhaps the whole ship—had been a casualty of the voyage. The sailors hoisted the timber upright and left it in that position, making a beacon of it to leave as a sign of their passage.

Barents sailed on, naming as he went. Capo Negro referred to the black terrain there, which reminded the sailors of a bar of slimy mud back home on the Zuiderzee. Admiralty Island, in honor of their commission, had uneven shores with dangerous shallows and couldn't be safely approached. They tried to re-create the new lands they sailed through in the image of their home country, to flatter funders and political leaders, and perhaps to imagine some future in which this northern world could become theirs.

At Williams Island—for William of Orange, not for Barents—the navigator again took the height of the sun. They had sailed more than two thousand miles from Amsterdam. It was July 7 when they came across more wood from a Russian ship that seemed to have made it as far up the coast as they had.

As the expedition left behind the realms of even indigenous Arctic peoples and made its way into lands unsuitable for human habitation, the calendar showed that the men were six weeks out of Amsterdam and less than two from Kildin. Again beset by fog, they entered a fjord by Williams Island and noticed movement. They saw something they'd never seen before. They recognized it as a bear, but it was a *white* bear. An enormous white bear swimming in the water.

Barents and his men jumped from the ship into their small boat. They had the idea that they'd try to take the bear back to the Netherlands with them and show it as a foreign wonder. Loading a musket, one of the sailors shot the bear full in the body. It reared up and began to swim off. The crew chased the animal, with several men rowing to close the distance while one managed to swing a lasso and rope its neck. Dragging the creature back toward them, the men pulled it along in their wake.

Barents had no prior experience with polar bears to prepare him. But he and his men would not be the only explorers to feel the impulse to capture a bear and bend it to their will. Even after the fierce nature and strength of polar bears was widely known throughout Europe, twentieth-century polar explorer Roald Amundsen asked the director of the Hamburg Zoo to transform polar bears into pack animals capable of pulling sleds on a polar expedition. Fellow explorer Fridtjof Nansen thought the plan worth attempting but remained dubious about its feasibility.[3] It seemed likely that polar bears hauling on real ice would soon sense their power in their home environment and revolt.

The optimism William Barents felt watching the first polar bear he'd ever seen was even shorter-lived. The lassoed animal roared and continued to fight, despite having taken a direct shot to the body. Though the Dutchmen didn't yet see their full danger, they quickly realized the creature's power. Revising their earlier fantasy, they decided it would be safer to kill and skin the animal than to keep it as a pet. While the bear fought to free itself, the sailors let the rope play out to hasten their quarry's exhaustion.

From time to time, Barents goaded the bear with his boat hook. After a while, the animal paddled up to the stern of the boat and gripped the wood with its forepaws. Thinking the creature wanted a break from the struggle, Barents called the men's attention to it, predicting "She will there reste her selfe."

The bear, perhaps sensing that it was in its own element, didn't rest on the ledge. Instead, the animal heaved itself up and climbed halfway into the boat. The sailors fled to the other end of the craft in terror. But the far end of the noose they'd put around the bear's neck caught under the rudder and leashed it in that vulnerable position, half in and half out of the boat, choking. The bear was trapped, not yet helpless but unable to close the distance with its attackers.

Realizing their position, one of the sailors made his way from the front of the boat and gored the animal with a half-pike. The bear collapsed into the water and dragged the boat for a time. Once the creature grew tired, they beat it to death. Hauling the animal into the boat, they began skinning it. The first polar bear the Dutchmen encountered would, in time, make its way to Amsterdam after all.

The sailors named the place Berenfort for their conquest. From there they went back to Williams Island, then on to the next bit of floating land in the tiny archipelago, where they left the ship once more. Taking their small boat to shore, they hiked through the cliffs and rocks up a hill to find a pair of large crosses on high. They weren't quite past the bounds of history after all—at least one Pomor ship had been here before them. They christened the site Cross Island, and moved on, mapping and naming as they went. Heading north, it was the last sign of human life they'd see.

Going to sea can be challenging enough with a plotted course and an endpoint in mind. Sailing day after day without a map into unknown territory is a completely different experience. It's impossible to know which kinds of shores, or what animals, wind, and weather may appear next. Any given tomorrow might have brought Barents or his men within sight of an open sea beckoning them to China. Or they might have worked their way north against ice and snow month after month until they came to the top of the world—the first humans to arrive at the North Pole. Or the sea might have risen up and swallowed them whole. They had to continue on without knowledge of their destination or how long it might take to get there.

The solstice marking the longest day had passed, but real darkness wouldn't return for months, when the tables would reverse and polar night would begin to creep in. That summer in full sunlight, standing on the shores of Nova Zembla, William Barents had no reason as of yet to ponder the possibility that he might ever be away from home long enough to discover what it felt like when light gave way to endless darkness.

They rounded a cape they named Nassau on July 10, and after that, the land curved away to the south. They continued along for the afternoon, avoiding a high sandbank that lay invisible far from shore. They'd come some two hundred miles north along the coastline of Nova Zembla. Were they merely heading into a bay, or had the ship already rounded the northern tip of Nova Zembla, proving it was one or more islands? In the far distance, they spied what appeared to be a new land far to the northeast.

If they were correct, they might discover a new northern land themselves. They headed toward it, sails billowing. But the wind blew too fiercely from the west, making safe navigation impossible, and they had to take in the topsail. When the wind blew even harder, they feared losing the masts, and struck every sail on the ship.

They lost sight entirely of the new land they'd seen, never realizing that it was likely only the curving Nova Zemblan coastline they'd spotted in the distance. They soon had bigger concerns. Late in the day, the waves rose higher, and between waves, the sea went hollow. With the water dropping and surging under the lurching ship, waves cracked overhead. They ran sixteen hours and some thirty miles without sails, limiting their ability to steer inside the fierce wind.

All night, the weather didn't relent. As the tempest drove them league after league over open water, now east, now south, the ship continued without sails. A vast wave rose up and swamped their small boat, tearing it away from where it had been secured to the ship and dragging it down toward the bottom of the sea, where the waters surged back over it and it vanished.

They rode the gale for hours more before the wind began to drop. The waves lessened, and fog appeared. The crew finally hoisted sails, and they made their way tentatively north and east into the mist. After they'd gone a few miles, the seafloor rose beneath them, and the plumb line dredged up black mud, suggesting that they were closing in on land. For the first time, they noticed drift ice floating on the sea.

During their time in the Arctic, Barents and his shipmates would see a staggering array of ice in various sizes and shapes. Pancake ice formed in circles that grew on the surface of the sea and gradually connected with one another as the air temperature dropped. Fast ice froze at the shore and worked its way out into open water. Tables of ice stretching like plateaus high above the boat could block a view entirely. At other times, fantastic towers and castles would rise in jagged Gothic architecture high above their heads. They were suddenly far from home.

On July 12, they spent most of a day sailing back to look for their boat. The next day, as the ship headed north and east, the relentless sun turned in its tight circle high overhead with no thought of touching the horizon. They encountered more ice. A sailor who'd been sent up the mast returned, reporting a flat white plane ahead as far as the eye could see. As they tacked a southward path around it, they caught sight of the Nova Zembla coastline again.

By July 14, Barents had the crew shift sails to follow the coast northward, but after several miles, ice surrounded the ship once more. Creeping nearer and nearer to an iceberg, they realized by the time they got close that it towered so far above them it was impossible to see over.

A hard wind blew. They navigated south and west again, then stumbled northward until once more, they hit ice. Each day became a numbing ritual with a jagged stitch sailed forward over the sea and then reversed in the face of white walls, with occasional sighting of the shores of Nova Zembla. It might take as long as half an hour to tack each time they had to turn: drawing in some sails to let the wind catch others, bracing the yards perpendicular to the mast by hauling on its spokes until they turned, unfurling the sails that would catch corners of the wind to go in a new direction, then belaying each line and making it taut. Surrounded by fog and sun and ice, Barents and his crew staggered forward inch by inch, sailing more than twelve hundred miles in the last half of July—as far as it would take to get back to the west coast of Norway—while gaining less than two hundred miles of forward progress in their quest to see if Nova Zembla continued on forever. The bitter sailing began to take a toll on the crew.

Barents sometimes measured the height of the noontime sun with his cross-staff. Less often, he used the mariner's astrolabe, rotating its dial, threading sunlight through two pinholes, and reading the sun's elevation. He calculated the ship's latitude as far north as 77 degrees, closer to the North Pole than any European had been known to sail. Trees had given way to shrubs, and shrubs to lichens. The vista of water and terrain visible from shipboard became more and more barren, fusing and transforming into a vast frigid desert.

Drifting white mountains moved in from the north as the sea itself began to freeze in chunks and flakes. Though the weather often remained fair, the presence of snow covering the land when it came in sight, not to mention the ice blocking their way, was a reminder that the height of summer had passed them by a month ago.

On July 29, the ship reached a place they named Ice Point, from which the land curved away to the east. Going ashore, the sailors found stones on the ground shining like gold. The story of Martin Frobisher's final expedition in 1578—which brought back fourteen hundred tons of stone that was determined to be nearly worthless—may have been on Barents's mind. There's no record of the sailors gathering any quantity of gold stone to bring with them back to Amsterdam.

Leaving Ice Point, the Dutchmen couldn't know at first that the land's sweeping turn to the east wasn't some bay or inlet. But in time, they would realize that they'd passed the northern tip of Nova Zembla. They continued north, to two rocky offshore outposts that they christened the Orange Islands, after William of Orange and his son Maurice. On July 31, they set foot on the larger of the two islands and found hundreds of sunbathing walruses, which they called seahorses.

The walruses lay huddled in clumps and piles along the beach. The Dutchmen headed to meet the animals, and watched with interest while the mothers swept their young into the water and protected them with oar-like flippers. After pushing their calves to safety, the adults moved to attack the invaders. The sailors had heard that walrus tusks could be as valuable as ivory. Seeing the animals out in the sun, the crew thought they appeared less coordinated and formidable on land. The men gathered hatchets and cutlasses and swept in, swinging their weapons. But the metal blades shattered on the animals' skin, and they couldn't manage to kill any of the creatures. They did, however, manage to crack off tusks, and fell back in order to carry them to the ship. There, they reasoned, they could get heavier guns and return, the better to kill all the walruses. Echoing explorers before and after them, the new arrivals' first impulse was to slaughter and plunder.

On the way to the ship, however, the sailors came across a lone polar bear asleep near the shore. They took the gun they did have at hand and shot the animal, chasing after it in their boat as the creature swam away. After killing the bear, they tied the carcass to a pike speared upright in the ice, thinking they would return for it when they came back with their guns to shoot the walruses.

But the wind rose up and broke icebergs into sudden motion all around them, driving the men back to the ship. They returned with only a few walrus tusks. Details of the hunt ended up in the official account of the voyage:

> *This sea-horse is a wonderfull strong monster of the sea, much bigger than an oxe, which keepes continually in the seas, hauing a skinne like a sea-calfe or seale, with very short hair, mouthed like a lyon, and many times they lie vpon the ice; they are hardly killed vnlesse you strike them iust upon the forehead.*[4]

Barents and his shipmates had been out from Kildin for five weeks, and at sea for two months. Though they were eating up their rations, months of provisions remained. Still, the endless tacking and slow progress made the sailors weary. The inexorable ice piling higher all around them was unlikely to diminish as fall arrived.

Ice had destroyed ships before their voyage, and would, in time, crush larger and sturdier vessels than any Barents sailed. In January 1915, after almost reaching his planned landing site on Antarctica, British explorer Ernest Shackleton would find his exquisitely crafted ship *The Endurance* frozen into the ice. The ship and crew drifted for months, traveling hundreds of miles west and nearly a thousand miles north, listening to the solid sea grind away at their hull. Sometimes it sounded like moaning; at other moments, the wood at the mercy of the ice cracked like pistol shots.[5]

Under the worst conditions, the sea could take a whole fleet. In 1871, a group of thirty-three whaling ships sailing between the far northern coast of Alaska and an expanding ice pack were destroyed when the wind shifted and trapped the vessels. The men were rescued, but the ships had to be abandoned where they sat, with their crews leaving them to be torn apart by the sea.

It wouldn't take much imagination for the sailors past Nova Zembla at the far end of the world to fear what would come next:

the thud of drifting chunks hitting the hull of the ship between decks, the watch for icebergs that might be large enough to stove or sink the ship, or the inevitable closure of any dim passage they might find now as winter came on. Even if they wanted to continue, it wasn't clear how they could. They'd surely end up spending the winter in a polar climate, keeping the ship as safe as they could in the ice and parceling out remaining rations, which wouldn't be enough to last until spring. The crew began to balk at the idea of sailing farther.

Mutiny against a captain was punishable by death, but crews did sometimes turn against ship's officers. After being trapped ashore on the eastern coast of South America during the winter of 1519, Ferdinand Magellan's men had been put on short rations as they tried to find a southern route to the Spice Islands for the Spanish crown. The next April, captains in the fleet led Magellan's men in rebellion against him.[6]

In May 1504, while marooned on Jamaica for a year during his final voyage, Christopher Columbus used lunar declination tables to predict eclipse and pretend to the indigenous Arawak population that he had divine knowledge. He also faced a mutiny led by the brothers Diego and Francisco de Porras—a mutiny put down only after a swordfight that one of the brothers lost to Columbus's own brother, Bartholomew.

Some captains whose sailors turned against them were greedy mercenaries. Others were simply tyrants. Nothing in the historical record suggests William Barents was either. Seeing no way forward with the sea as it was, Barents agreed to give up hope of circling Nova Zembla and following its eastern coastline down to meet the *Mercury* and the *Swan*. On August 1, 1594, they turned the ship around.

The men worked their way back from the walruses of the

Orange Islands to Ice Point, and then miles along the coast, where a group of low dark hills stood like the mirage of some lost country huts. Sailing away from land, they passed the wrecked Pomor vessels they'd seen on their way north. Freed of the paralyzing menace of the ice, the return was easier than the voyage out. By August 8, just a week after they'd turned around, they had traveled four hundred miles south along the sweeping line of Nova Zembla's western coast, a long, narrow, arc of land curved like the neck of a goose. (The fat southern portion of Nova Zembla would in time be called Goose Land.)

The sailors came to an island of black earth in the fog. Unable to see any safe distance beyond the bow of the ship, they headed out to sea to be sure to keep clear of the shore. When the sun returned, they sailed back and saw that they'd arrived at the inlet of Kostin Shar toward the southern end of the island mapped a decade before by Olivier Brunel on a mission financed by Dutch merchants.

Barents had wanted to hold to the shore, but the ice that had stymied them in the north made a return visit. The crew headed west into the sea until they worked their way clear of it. Sailing close to land again, they spied more black rock, and a cross near the shore. They knew they were nearing the southern tip of Nova Zembla and had returned to lands rumored and described on prior expeditions. The men couldn't be sure that no other travelers, traders, or Sami or Nenets herders were in the vicinity.

Taking the small boat ashore, they saw signs of a human presence, but the residents seemed to have fled recently, perhaps from Barents and his crew. The Dutchmen explored the area and identified the carcass of a broken Russian *ladya* with its long keel, along with six buried sacks of rye meal, a stray bullet for a long gun, a second cross, and several staves for barrels. In all, they found three wooden houses built in a northern style. Nearby lay five or six coffins filled with bones and stones.

Seeing no living thing, they called the site Mealhaven and sailed on. During six more days of fog and sun, they worked their way to the bottom of Nova Zembla and came to the islands of Matfloe and Delgoy, where, against any reasonable odds, they spotted Cornelis Nay and the rest of the fleet.

Greeting their mates from Amsterdam with gunpowder and celebratory explosions, the crews of the *Swan* and the *Mercury* at first thought that Barents had circumnavigated Nova Zembla as planned, sailing up its western shore, over the top, and down its eastern coastline all the way to Vaigach Island. Before long, this impression was corrected, to everyone's disappointment, and the captains told Barents and his men what they had seen while sailing near Vaigach Island during their six weeks apart.

The *Mercury* and *Swan* had made their way from Kildin Island into the open sea and had found the ship surrounded by whales and black-throated arctic loons. Within a week, they'd encountered imposing sections of hard, flat ice on the surface of the water, as well as icebergs twenty feet high or taller floating with the current. Making their way between the floating islands, ice surrounded them on all sides. They could see little water and no land, with no apparent exit. After some three miles of sailing in the blank labyrinth, seeing mist rise in the distance, they thought they might be approaching land. "Sea dogs"—seals—littered the tops of icebergs, while geese flew overhead. Eventually, they'd made their way out of the maze of ice, sailed back to the sea, and continued heading southeast.

Two days later, they caught sight of snow-topped high terrain that stretched level and even, though fog blocked any clear view. Later the same the day, they came to an island with a hillock near the shore at the top of which sat a cross. Soon after, they stumbled again into a terrifying seascape of icebergs that seemed to change moment to moment, some with grim caves, and others big enough for the sound of water crashing against them to sound like waves battering the shore—a fate that would be disastrous for their ships. They had little idea of where they were. They saw plants, feathers, roots, bark, branches, and the wood of trees like relics of some shattered landscape. Small birds floated overhead as if looking for land, and two that might have been swans winged their way north-east in the distance.

Ice blocked their route again and again, covering the water so snugly and seamlessly that they appeared to be lodged in a conti-nent of ice that stretched away into the distance—even from the bird's-eye view atop the mast. There was no longer any water to sail in, only ice.

When they finally managed to get free, they turned to avoid floating icebergs and to watch for land, which was hard to match with the known geography of the region, due to mist and fog obscuring everything. When they sent out their small yacht, which could navigate the glaciers with less danger, it found a harbor safe from the threat of ice, and the ships made their way to safety.

The *Swan* and *Mercury* soon had company. The crew of a Rus-sian *ladya* plying the coast told them that they were at least a day's sail from where they thought themselves to be, having covered more than half of the distance between Kildin and Vaigach Island, south of Nova Zembla. The captain of the Russian ship made a sketch of the coastline for them, but it was a collection of land-marks with no indication of latitude. The captain said he had no personal knowledge of Vaigach Island but had heard rumors that the passage between Vaigach and the Russian mainland was nar-row and blocked by ice. On the other side, he said, lay a southern sea known as the Warm Sea, as opposed to the northern sea, known as the Cold Sea.

They ran into two more Russian ships, which gave them similar news of the strait between Vaigach Island and the mainland, although one added that it would be possible to get through, if it weren't for the vast numbers of whales and walruses there. They'd also have difficulty, the Russians added, with the sandbanks, the rocks, and the waves driving their ships onto the shores of the passage. As if that weren't enough, these new acquaintances reported that the grand duke or tsar had recently sent three ships to navigate Vaigach Strait, only to lose all three, along with several sailors, in the ice.

Seeing more Russians on shore, Nay's men brought them aboard and learned that the coast was peopled with foreign hunters and trappers, but some had hidden themselves at the sight of the Dutch ships. Told that there was nothing to fear, those who were hidden came out, later inviting the ships' crews to join them on hunting and fishing expeditions.

The Dutchmen had seen footprints of bears, as well as wild geese and duck on the wing, but the living things that most distracted them on windless days were the pervasive gnats. More exciting was the arrival of whales surrounding the ship in mid-July. Without harpoons, they couldn't easily capture or kill the creatures. Instead, the men drove them into the shallows to wear them out. They finally managed to finish one whale off after a long chase by stabbing it in the back. The dying whale fought as it bled out, painting the sea red. They dragged it to shore, chopped it to pieces, and dumped the blubber into barrels for oil. While they were cutting it up, another whale approached the shore where they worked, but they didn't try to kill it, having no room left to bring aboard blubber from a second creature.

Russian fishermen on their way to the Pechora River gave them fresh fish and reassured them that the icebergs that had begun to surround them would melt in the next several days. Sometimes the frost and mist were such that the sun was blotted out; on other days it hung in the sky, ominous and crimson. The icebergs loomed larger, as big as floating islands, but they were porous and prone to breaking into pieces and vanishing into the depths before the sailors' eyes.

On July 22, they caught sight of land, which they mistakenly believed to be connected to Nova Zembla itself. The ships sat trapped in fog, but when the sky cleared, they took a reading and found they'd reached a latitude of 70 degrees 20 minutes. They'd come to Vaigach Island. As with Barents finding the northern tip of Nova Zembla a week later and hundreds of miles north of them, they hadn't found the Arctic to be hospitable or helpful, but they'd managed to reach their destination.

Again they encountered driftwood, trunks, roots, and branches of trees entirely covering the surface of the dark water. Their soundings revealed a shallow waterway over pebbles, then more dangerous stones and boulders. They came to a place with twin crosses; believing it inhabited, they rowed their small boat in to look around. They wandered near the coast and spotted a man. They hurried to catch up and surprised him. He didn't say much and clearly didn't want to talk to them, but he understood a little Russian. They advanced on him to seize him. He was frightened and managed to escape.

Jan van Linschoten, who kept a diary, said that the escapee reminded him of the inhabitants of Kildin Island whom they'd left behind when the ships had split to follow their separate routes. Having heard from the Russians that the coast along Vaigach was inhabited by groups of these people, the men aboard van Linschoten's ship became even more certain that they were navigating the strait as planned.

They had sailed a lusher landscape than the one seen by William Barents—at times a beautiful terrain of hills and mountains, with whole fields covered in every color of flower, whose scents reached them. Yet in other places along the coast, vegetation was sparse, and there were no forests. They'd seen walruses and reindeer, as well as scat from a number of animals they couldn't identify, but few birds other than an occasional passing chaffinch, swallow, or seagull. They far more often saw an entire tree, from roots to crown, floating in the water, or even thrown whole onto high land, the magnificent corpse of a storm.

Still above the Arctic Circle, they were nonetheless far enough south that, on the night of July 23, the sun dipped below the horizon before rising in almost the same corner of the sky a short time later—the first time in more than a month that they'd seen a sunset. They could tell again by the presence of crosses that Russians had preceded them in that place, but by the character of the landscape to the south, and the currents carrying ice toward them, they were now fully convinced that they'd been sailing the waters between the mainland and Vaigach Island. As they pressed on, the dark water turned blue and salty, recalling the open ocean.

On July 26, though the sky was clear, the weather was cold and a strong wind began to blow. They dropped anchor to venture ashore and found reindeer horns on skulls that had been gnawed to the bone. Snow and hail drove in hard under dark skies, sending them back to the ship to wait out the weather. Van Linschoten brought with him a walrus head likewise chewed clean, for further study.

Icebergs began to come in thick, some headed straight for the vessel. As the ship was pulled toward a strong current, the speed of the flow didn't leave time for escape maneuvers. One near miss struck the northern coastline instead and ricocheted back at them too quickly for the crew to hoist anchor and flee. To try to give the ship room to move with the ice, they ran out the anchor cable. But it snapped like a match. They lost their anchor and were swept deep into the current amid the icebergs.

A ship of any size carried more than one anchor for just such occasions. After hoisting the mizzen sail, they managed to get clear of the ice and cast out a second anchor. Again the ice rose all around them. As they fed out line to drift and avoid one threat, icebergs struck the other side of the ship like a collision with rocks. The anchor cable got twisted under the floating mountains of ice. The current dragged the ship again.

Then the arms of the second anchor broke off and remained trapped in the seafloor, leaving the sailors holding only the anchor rod on the end of the line. They were adrift once more. They set the sails up to tack against the ice. Making their way to a calm strip on the north shore, they found enough shelter from the storm and good ground to drop a third anchor within cannon-range of land. They recorded a latitude of 69 degrees 43 minutes and paused to christen the unsettling passage in which they found themselves the Strait of Nassau.

Sailors went out in the small boat and returned just as night came in. They came back carrying the bower—the first of the two anchors that had been lost—trailing the strip of cable that had snapped in the gale. The sky had gone to storm again, and the east wind drove a damp cold into them without mercy.

As they set out near noon on the next day, they spied a southern coastline with raised pieces of wood, and at first took them for the Russian crosses they'd spotted before. Coming closer, they saw instead hundreds of wooden idols of all sizes on a stretch of beach choked with reindeer antlers. The idols were male and female, adult and child, and all had their carved faces turned to the east. The sailors assumed they'd stumbled onto some sacred or sacrificial land. Because some were newer and others were riddled with rot or worms, van Linschoten came to think that perhaps each idol honored a dead native. Some were cut totem-style, with several faces stacked above one another on a single piece of wood. Nearby stretchers with hewn feet seemed to be some kind of bier on which to carry the idols. Though it was clear that people came there regularly, the men could see no sign of current inhabitants. Nearby, more driftwood and pieces of rotten walrus carcasses floated in the water.

The land itself was rich and green with small basins and lakes filled with snowmelt, which could be brought down in barrels as needed. Despite the sea rubbing right against rocky shores, this spit of land might make a good stop for refilling stores of water on future voyages.

But every good piece of news was overshadowed by grimmer events. The *Swan* and *Mercury* were pinned in for two days by storms and the constant deadly threat of ice. On the morning of the third day, July 29, they spotted an iceberg at least half a mile long moving through the channel. If the current nudged it sideways, it could close the strait entirely. Up to that point, Admiral Nay had stayed aboard the *Swan*, but seeing the vast approaching danger and knowing it could smash the ships, he headed to shore with the crew. The visible parts of the behemoth blocked their view, filling the sailors with awe as it approached. It pulled alongside them in all its grandeur, then moved on without turning, and passed them by. They began to wonder where else such massive islands of ice could come from but the open sea.

The sailors went ashore the same day and found a small cabin, along with more elegantly worked idols than those they'd seen on the shore. In the distance, they spotted three reindeer pulling a man on a sled. They called to him, with an idea that they might capture the newcomer. But as the crew moved in to take him, he raised a cry. Dozens of similar people appeared, likewise on sleds pulled by reindeer, surrounding the sailors. They barely escaped to their small boat, which carried them back to the ship as arrows chased them out of range. It became apparent that the next time they wanted to ask questions about what lay on the other side of the strait, they'd need to make a bid for friendship rather than abduction.

To the east, the shores of the mainland and southern coast of Vaigach Island veered away from each other. A broad expanse of open water stretched before them. The men of the *Swan* and *Mercury* became more certain that they'd cleared the strait. Three days later, they saw the frozen north shore of Vaigach Island rolling up and away toward Nova Zembla. But just as they thought they were free, winds from the east blew calved pieces of glaciers into the strait, as if to deliberately hinder them. They'd reached the very end of July. Though summer still made itself felt, it was hard to imagine navigable seas at any higher latitude.

In reality, William Barents sat that day at the far end of Nova Zembla, having managed to sail hundreds of miles farther north than the rest of the fleet. But almost in that moment, his own misgivings and the crew's trepidation were convincing him to turn back.

Going ashore to scout the terrain, van Linschoten and his shipmates saw another group of herders driving reindeer sleds. Changing their approach, they sent out their Russian interpreter with one other man, both unarmed, "so that we would not frighten the barbarians." The men approached, watching warily for any sign of ambush from the ship or elsewhere. After exchanging greetings, the sailors gave their new acquaintances bread and cheese, which was eaten. A half-dozen Europeans joined those already on shore, where the archers allowed the newcomers to handle their bows. But they wouldn't turn over their arrows for inspection. The present sea, they learned, was a small one, and if the men continued eastward, they would find a vast body of water there. The Dutchmen got the intelligence they wanted regarding navigation. The Nenets they spoke to claimed that, after several more days, the ice would subside, and the strait would be clear for six weeks before its return.

Asked if all this land was the territory of the grand duke of Moscow, the Nenets man replied that they'd never heard of the grand duke. They knew of Russians who came to hunt and trade with them, but the land wasn't settled by any permanent inhabitants. They likewise didn't use the name Vaigach, but had their own names for the land and the water.

A willingness to help the Dutchmen didn't endear the Nenets to van Linschoten. He noted that the reindeer herders wore their mittens stitched to their sleeves and their hats attached to their jackets, like crude Dutch peasants. Their sleds were shaped more like chariots than those he'd seen the Sami use on Kildin Island. He found these people small and deformed, writing that some resembled monkeys or monsters. He seemed suspicious of their beardless faces. (European gentlemen wore beards.) All in all, van Linschoten surmised, they were a "miserable, defiant" people with little of value to offer, belligerent, and hard to discipline. If some future expedition were organized merely to trade with them, he suggested that "the game would not be worth the candle."

Other travelers had a better impression of the Sami and Nenets, but van Linschoten was vitriolic in his assessment, reflecting the strain of the new Dutch nation that would come to engage in the slave trade in the Far East and the Americas. Taking leave of their hosts, the Dutchmen went aboard their ships and sounded a trumpet in salute, sending their new acquaintances fleeing in fear. In time, it was understood that no threat was intended, at which point everyone waved their goodbyes.

July turned to August as they sailed along the coast, marking a safe harbor with a buoy and naming an island after a patron of the expedition. The sailors made their way out of Vaigach Strait and into open water, which they named the New North Sea. (In time, it would be known as the Kara Sea.) They felt sure that they didn't have far to go, and that the waters of this new sea would stretch all the way to the shores of China and Japan. The crew had no way of knowing that their ships still lay thousands of miles from the East Asian coastline.

Once more, they tacked in the face of icebergs that crowded together and loomed like mountains into the vault of the sky. But the water near shore now stood so clear that the men could see crayfish scuttling along the floor of the sea. They were grateful that no storm was brewing, but when the storms did come, it was simpler to ride with the current than try to navigate the terrifying bombardment of ice.

The following day they spotted a huddle of walruses, which van Linschoten noted might "better be called elephants of the sea rather than morses or sea-horses." Like Barents's men, the sailors who set out with van Linschoten knew that the tusks of the creature were as valuable as ivory to the Russians—and their response to seeing the creatures was almost the same. Some of the commander's men began shooting, and, having wounded one of the animals, thought they'd chase it in the small boat. They went out after the injured walrus and speared it through the body. But no matter how hard they struck at it with their bladed instruments, its hide bent their hatchets. The creature counterattacked against the small yacht with its tusks and threatened to tip it over. After an hour and a half of grueling exchanges, they gave up and left the animal with blood running out of its nostrils across the surface of the sea.

The sailors came upon an island with a small bay, where they anchored between stones so weathered by ice and snow that the terrain brought to mind the Latin phrase *urit frigus,* "the cold burns." They explored the coast and discovered a kind of rock crystal that resembled diamonds. The minerals seemed already cut and polished, and the crew began collecting pieces of it, though some turned out to be too fragile to cull. It seemed a wonder that such things could even be found so far north. Whatever their value might turn out to be, they'd found a small, dependable safe harbor on the far side of Vaigach Island. They named it Staaten Eylandt—meaning "States Island"—in honor of the States General, the Dutch parliament.

They made their way from States Island back to the mainland and explored the coast, where van Linschoten stole one of the Nenets carved idols. The men kept girding themselves to head out into the sea to seek some greater open passage, as soon as the ice cleared. And on August 9, the two ships finally made a stab eastward, sailing the shores of what seemed like open ocean, above a floor too deep to be measured with their plumb line.

The vessels traced the coastline from rocky shores to a river delta, where crew members thought they saw a ship full-rigged with sails, only to realize it was an iceberg. Another time, they sent a man up the mast who sighted humans and wild animals walking the land, but on closer examination, no living creatures were there. They sailed on in unnerving isolation. The sailors began to realize that fog and mist over water, as with sun and sand at desert latitudes, could inspire mirages.

By August 11, they'd journeyed eastward on the coast for more than a week and believed they'd sailed past the river where explorer

Olivier Brunel had met his end. They thought they'd discovered two more rivers farther along, which they named Mercury and Swan, after their ships. According to their speculative maps, a long cape lay at the far northeastern corner of the continent. Around the corner of that cape, they mistakenly believed, the shoreline descended toward China.

Seeing the coastline arch to the north, they thought they'd arrived at the fabled cape. If so, there would be nothing of the route left to discover. They no longer doubted that they had found a navigable route to the Far East. The window for them to sail east and still return to Vaigach in time to meet up with Barents and his boat was closing. Delighted with their progress, they set their sails for a westward journey and turned for home.

The crew unwound the route they had sailed, scouting more coastline as they slowly made their way back toward Vaigach Island. Time after time, but always at a distance, they spotted whales half out of the sea, jetting plumes of water into the sky. The men took the sightings as yet more proof that they'd found open ocean.

Coming into mid-August, nighttime meant a return to real twilight, even a pale view of the moon and stars. When the winds rose at night, they couldn't anchor safely away from icebergs or land and be sure the ship would stay whole. Nor could they unfurl their sails and steer in night and fog. So sailors hoisted the anchor and let the ship slip without sails once more into the current, keeping watch and groping their way along.

In daylight, they fared better, but the wind blew hard. Navigating shallow waters as they followed the shoreline posed risks, too. Steering between two small islands, they tacked into waters that grew shallower and shallower, constantly measuring the depth of the seafloor, crossing through just submerged rings of rock with only dexterity and the brutal wind to pull them through. When the wind blew toward land, there was always the chance of smashing to pieces on a lee shore. They moved past a tidal wave crashing in whitecaps for nearly a mile, after which the men came upon sandbars sitting in their path. They couldn't go forward in safety without water deep enough to float the ships, yet the prospect of what they might find in trying to retrace their precarious steps was terrifying.

The vessels reversed direction and began to thread their way back through the maze of sand and rock near Vaigach. As they did, two buildings came into view atop one of the tiny, uncharted islands along the shore. Drawing closer made it apparent that they weren't looking at cabins at all but were instead seeing the topsail and topgallant of a ship. To their shock and joy, they recognized that it was the ship from Amsterdam. In the far north, thousands of miles from home, they'd stumbled across William Barents.

Despite their excitement over the reunion of the small fleet, the sailors weren't yet out of danger. Measuring the breakers coming in, they realized their ship had at one point been sitting in less than twenty feet of water. The wind was high, but if they had sailed the same stretch of shore at night, in fog, or during a storm, it could have meant an end to the ship.

They worked their way to the other side of the island, and the commander sent men out in his small boat to pick up Barents and row him back. Sitting down together, Barents told Nay his story of pressing northward above 77 degrees of latitude, more than five hundred miles from where they sat at anchor, only to be halted by ice. Van Linschoten later noted, somewhat acerbically, that Barents hadn't found the far northern passage over Nova Zembla to China, and so had come south of Vaigach to scout locations and follow the better route—the route that the commander of the fleet had already discovered in the meantime.

The night of August 15, the reunited ships tacked their way

offshore from the islands, and in the morning, looked for a harbor to wait out the bad weather. Exploring island shores two days later, they found bones from haddock, whiting, and cod, as well as the wreck of a Russian ship, and a whole tree some sixty feet from roots to crown and three feet across. They kept an eye out for forests or perhaps a stand of trees but saw nothing, not even plants. The sailors did, however, take advantage of the presence of swans, wild geese, ducks, and other sea fowl, capturing fledglings and shooting older birds. They had two live falcons taken from the coast of Vaigach Island, which they hoped to carry back with them as a gift to supporters. In the meantime, they named the island Maurice Island, after their leader and benefactor who had backed the expedition, and christened the coastline of the mainland below it New Holland.

On the twentieth, a Saturday, they took an easy west-north-westerly course until evening, when the wind began to pick up. Thinking they saw land to the southeast, they soon realized it was only fog. That night the wind rose again, now with real force, and sent the ships flying in the driving rain. They set sail to follow the wind west, but were running blind over the wet and overcast seascape, moving too quickly to run out a line to the seafloor and no way to know if they were in danger.

Without warning, the *Swan* struck land. As the hull scraped along the shore, Admiral Nay called out to warn the rest of the fleet. But the wind dragged the *Mercury* in too fast. Before it could turn, it ran aground hard enough to knock it off balance. Bringing up the rear, Barents saw the fate of the other ships, and was able to change course, avoiding the coast. The *Swan* came off the shore without difficulty and remained seaworthy. But the *Mercury* didn't come out on the next wave, or the one after it. The ship tilted miserably, wallowing on land.

If a vessel ran aground hard enough to breach the hull, the crew would have to work the pumps to keep from sinking. Pumps had been an integral part of seacraft for nearly two thousand years by then, with screw-pumps, buckets, or tubes that drew water out of the hold and up to the level of the main deck. But from buckets to treadmills and hollow tree trunks, bilge pumps could only carry so much water. And if a ship had a massive hole, and weather conditions remained precarious, a ship could be lost.

Luckily, the *Mercury*'s hull also remained sound. But it would take more than twenty tries to heave the vessel, groaning in the rain and wind, off the shore and set it floating once more. Only the smoothness of the shoreline had saved them. If the *Mercury* had crashed onto a cliff or ledge, it might have been impossible to break free.

They sailed back past Kildin Island, and along the Norwegian coast once more. With the wind against them, they waited out storms at Wardhuys, where a Danish customs officer demanded their passports and payment. After showing him a letter in Latin indicating that they were an expedition chartered by gentlemen, and giving him a handsome tip, which he said wasn't necessary but didn't return, they became friends. He seemed particularly pleased when they recounted that their expedition had clearly failed, and implied they had no intention of returning to explore a northeast passage again.

Summer slipped away with daylight as the fleet headed farther south. By the time they were back in the North Sea, they were approaching the fall equinox. They'd been out for three months and ten days, and Barents's ship had sailed farther north of the continent than any European expedition in history. Admiral Nay and van Linschoten had crossed Vaigach Strait and sailed deep into the sea east of it. They now knew that Nova Zembla was almost certainly one or more islands—islands that could be circumnavigated—and not some impassable polar continent. They held close the promise of a direct route between Vaigach Strait and the northern corner of the mainland, with warmer latitudes and coastal sailing all that remained before they might arrive safely in China.

At two in the afternoon on September 16, 1594, they caught sight of Texel Island and Huysduynen past Vlieland, and sailed by them at high tide, coming back into Dutch waters. The *Swan* peeled away toward Zeeland, leaving the *Mercury* and the ship from Amsterdam to enter the Zuiderzee. The *Mercury* headed to Enkhuizen, while William Barents sailed on, south and west, coiling back into the nested fist of Amsterdam harbor. Combining the split routes taken by Barents and Admiral Nay, the fleet had sailed more than three thousand miles north and east, going deep into unknown waters in both directions. They'd found two possible routes to China, and every man aboard had come home again.

CHAPTER THREE

Death in the Arctic

Carrying the corpse of a walrus, the skin of a polar bear, and two live falcons, the fleet sailed back into harbor on the day of the Amsterdam city fair. Barents and company met with a jubilant reception from the burgers of Amsterdam. Jan van Linschoten, the merchants' representative, continued on to the Hague, where he reported to Prince Maurice the discovery of a navigable, if treacherous, northeast route to China.

Prince Maurice and the States General—the advisory body of delegates from each province—authorized a second expedition, one intended to establish trading partners. All the sponsors of the first voyage—Amsterdam, Enkhuizen, and Zeeland—felt confident enough to risk their money supplying ships for another attempt on the Far East. This time, they'd stake generous amounts of their merchandise as well. Seven ships were chartered to sail out: two from each city, and another from Rotterdam. While a larger fleet ran a greater risk of attracting pirates, if it could stay together, a convoy would also be better able to defend itself and carry a wide range of goods to tempt eastern partners into trade.

Barents's first northern foray had brought impressive results. By cresting northern Nova Zembla, he transformed knowledge of the polar regions and undercut the belief in a vast polar continent. Geographer Petrus Plancius, who championed Barents and had tremendous influence, still felt the northern route over Nova Zembla to be the most promising.

But if a warm polar sea existed, Barents hadn't yet found the path to it. Political influence came into play as the fleet's backers debated how to proceed. As "supercargo" aboard the Enkhuizen ship—the person delegated to represent the merchants' interests—van Linschoten had the ear of all parties involved. He pressed for the route under Nova Zembla through the strait at Vaigach as the true path to riches and glory for the new republic.

In the end, the fleet was commanded to sail through the strait and try to reach a king or other Chinese official with whom they could negotiate trade relations. Six ships would carry goods to trade. The seventh, a pinnace—a small boat normally used to ferry people to land or from one ship to another—would confirm safe passage of the convoy into the open waters of the Far East and then return directly to the Netherlands to deliver word of their success.

As before, geographer Plancius helped to draw up plans for the voyage, laying out the territory believed to lie between Moscow and the eastern shores of China. After the known lands controlled by the grand duke of Moscow came the Tartary Coast, a vague geographic idea encompassing modern Siberia. Past that, terrain curved to the north before dropping steeply southward. There, they hoped, they might find the lands that they'd heard about—and perhaps others that were entirely unknown to them.

The *Swan* and the *Mercury* had made progress on their voyage but had mapped only a small part of the coastline on the other side of Vaigach. There were those who still insisted the region couldn't be sailed. This was a strange argument, given that the route south of Nova Zembla was known to be littered with Russian crosses from Pomor expeditions and inhabited several months a year by indigenous reindeer herders. But the failure of prior English and Dutch efforts, as well as the harrowing ice that had threatened Barents and van Linschoten alike, led to controversy over whether any northern route could be found. If the Dutch were to transform their country into a world power, they needed to capture key routes while trade with the east was still being established.

Amsterdam intended to hedge its bets. The Dutch authorized a second, separate voyage slated to leave in spring of 1595—one that would take a southern route. By this time, Jan van Linschoten's accounts of his time in Goa working for the Portuguese—the maps he copied, the records he took, and a vast bundle of intelligence collected through his earlier position overseas—had been assembled for print by an Amsterdam publisher experienced in marketing books on travel and navigation.

Van Linschoten used his own experiences and accounts from others to describe countless new regions and their idiosyncratic features, from the use of elephants in Burma to freakish animals to tea drinking in Japan. With the publication of his massive project,

van Linschoten's name would become known among Dutch political leaders and merchants. His writing about his travels included
routes and landing sites that had taken decades for the Portuguese
to find and establish.

New Dutch knowledge of Spanish and Portuguese voyages
into the Southern Hemisphere made it possible to imagine a way
to defy these rivals—to steal proven routes and trade partners from
under their noses. To that end, van Linschoten's intelligence was
consulted to determine a course. The plan was for another fleet to
sail south around Africa while Barents and van Linschoten made
their second northern voyage.

Equipping a small fleet for reconnaissance was one thing. Preparing a trade convoy was another. Merchants sent goods to the
respective ports, and ships were outfitted with a year and a half
of provisions, double the usual personnel, and ammunition in the
event of complications.

Cornelis Nay returned as admiral of the fleet heading north,
weighing anchor from Zeeland aboard the *Griffin*, a ship capable of
carrying about twenty-two tons of cargo. Enkhuizen and Amsterdam sent the smaller *Hope* and the *Greyhound*, respectively, both
newly built vessels roughly half the size of the *Griffin* but decked
out for fighting if an enemy appeared. All three ships would also
bring a companion yacht.

Barents would sail as "captain and coxswain" and "first pilot"
for the expedition.[1] Joining Barents in the Amsterdam ships was
Gerrit de Veer, from an established Amsterdam family. His father
was a notary, and his brother a legal scholar. De Veer would record
the journey for history, as would van Linschoten. Van Linschoten
would sail again—this time joined by a young sailor still in his
twenties named Jacob van Heemskerck, who would serve as supercargo, and Jan Cornelis Rijp, who would captain one of the ships.

Each man was sworn in prior to departure, and on June 18, the Amsterdam ships left the harbor for Texel to join the fleet.

Their homeland, however, wasn't ready to let them go yet. They were delayed two more weeks on the island due to complex preparations that threatened to ruin everything. The men who'd sailed the prior year surely remembered the ice they'd encountered, and the narrow, uncertain window for safe travel. But despair couldn't move men and commerce any more than it could shift heaven and earth. By the time they all set out from the southern end of Texel, it was July 2, nearly a month later than they had sailed north from the same island the prior year.

They were charged with making arrangements for trading vessels to harbor in foreign lands, and to conduct "pious, faithful and sincere steady dealings, traffic and navigation for common prosperity." They were expected to model the behavior they hoped to meet with themselves. They had food and ships and cargo and men to protect it all. They'd been provided everything they needed to make history.

After their delayed launch, the first weeks of the voyage led them through well-known territory, offering few surprises. But shifting weather and the necessity of so many ships staying together even when it caused delays didn't work in their favor. They had days of rain and storms, and a rough wind from the north blocked their progress.

The wind was a sailor's best friend—making it possible to move unbelievable amounts of cargo as if it were weightless. But it could also wreck and destroy ships. A vessel trapped between a hard wind and an unforgiving shore was a captain's worst nightmare, and storms at sea could unleash danger in an instant, especially if no known harbors lay nearby.

Ships weren't the only thing at risk. A tossing deck at night or in poor visibility had led to the disappearance of countless sailors across the centuries. Most European sailors, even experienced ones, often had no idea how to swim and could be superstitious about going in the sea—with good reason. Any sailor going overboard unnoticed could easily be dead in short order. Even if he were sighted, unless he were thrown a line within reach, a large ship skating on a good wind might not double back in time to rescue even a proficient swimmer. To be at sea without a vessel, even near shore, was often a death sentence.

After the late start out of Texel, the ship from Amsterdam was delayed by its pinnace, forcing it to catch up to the fleet. On July 21, they came upon a whale that they nearly rammed head-on as it slept. But it woke in time and cleared the way for the ship to pass by.

Two weeks later, two miles from shore, they'd almost completed the first leg of their northern route up the Norwegian coast and were making steady progress toward Nordkapp, where they would change course and head east. The fleet was sailing as a group when, without warning, one of the ships ran hard onto a rock at high speed, splintering the prow. The captain called out in dismay and warning.

The crew flocked to the pumps to start drawing out the water that would flood in after any rupture of the hull. The other crews, hearing the warning, jumped into action and steered clear of the damaged vessel. Sailors running down to check the cargo hold found that somehow the rocks hadn't punched entirely through. The wind turned the ship a little, and the ship's crew managed to creak and heave it away from the jagged shore. Their hull intact, they sailed on.

Between the bad weather and the encroaching rocks, they might have earned a reprieve. But after rain days later, fog descended. The seas began to rise, and clouds rolled in from the southeast.

Amid pitching waves and wind, the fleet continued to sail. The *Greyhound*, with William Barents aboard, sped along at the back of the convoy but suddenly came careening from behind.

The ship was thrown forward by the storm. Sailors called out for Barents's vessel to tack away, but neither time nor weather allowed a change in course. The ship drove in again, its prow plowing into the starboard side of the *Hope*. The grinding of the ships against one another and the sound of splitting wood seemed to announce the destruction of both vessels. The nose of the *Greyhound* plowed over the deck of the *Hope*.

The horror was compounded when the mizzenmast on the *Hope*—the rear of its three tall masts—came crashing down at the stern among the waves and the splintered wood, smashing everything in its path. The *Greyhound*'s mizzenmast soon fell, too. As the force of impact turned the ships, the built-up wood of the beakhead at the prow of the *Hope* was sheared off.

Two masts coming down between ships that had run afoul of each other in a storm meant double the risk of punching a hole in either or both hulls. Yet the wind and sea didn't smash the two ships together again. Instead, the ricochet from their impact dragged them apart. While the wind raged on, each crew took stock of the situation in fear, fury, and blame. The damage to both ships was extensive.

But all was not lost. The structures below the waterline weren't breached. Though the mizzenmasts had fallen, the spars hadn't been lost in the storm or cut adrift deliberately by a captain worried about the damage they might do. Still, the news was horrifying: four men who fell overboard during the mayhem had drowned.

The sailors moved the debris and secured the masts. If they faced a grim scene and couldn't bring back their mates, they at least had a solution at hand to deal with the rest of the disaster.

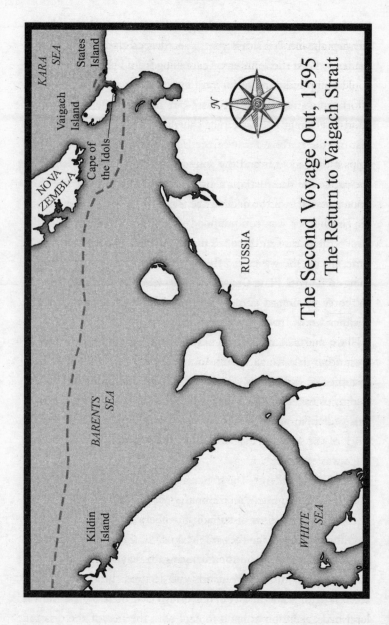

KARA SEA

States Island

Vaigach Island

Cape of the Idols

NOVA ZEMBLA

RUSSIA

BARENTS SEA

The Second Voyage Out, 1595
The Return to Vaigach Strait

Kildin Island

WHITE SEA

In a feat of engineering that was part of the skilled repertoire of any captain and ship's carpenter, the crews used the main masts as cranes to hoist their junior neighbors back into place, where they could be bound and braced into their former position. Any such effort ran a small but real risk of capsizing the vessel, and many hands were required to manage the project. Though it took the rest of the day, and the weather still refused to cooperate, both ships managed to raise their mizzenmasts again. The damage to the prow was addressed as much as it could be in short order on open water; the rest would have to wait for a time at anchor or better weather.

The following day they spied a Dutch ship gaining on them. In time, the vessel, out of Enkhuizen like the *Hope*, caught up, and they exchanged messages with the captain. He was on his way to the White Sea to trade, which didn't affect them at all. But he let them know that he'd also set out from Texel leaving the Netherlands, and had left two weeks after the first ships had set out. Despite his later departure, he'd managed to catch them. They'd lost two full weeks to foul weather and the inconveniences of sailing together. They'd now been out more than a month and were still on the Norwegian coast.

The ship joined their fleet. They soon caught sight of the *Iron Pig* out of Amsterdam as well, and sailed on as a nine-ship company. They rounded Nordkinn as a group, cresting the European continent once more. After four days of travel together, the two latest members of the convoy split off to make their own way south toward the White Sea. On August 13, Barents and the original fleet headed east and southeast toward Vaigach Island.

Four days later, they encountered the first bright sun they'd seen in days. But near noon, they spotted a familiar ghost on the landscape, as ice began its creeping return. One iceberg stretched in a flat, broad plane as far north as they could see, until sailors went up the mast and imagined perhaps that they caught a glimpse of open water beyond it in the distance.

They sailed southward along its edge during the rest of the day and all night, as if along a coastline, but mile after mile, the iceberg—if it was only an iceberg—had no end. Their ships lay perhaps thirty or forty miles from the shores of Nova Zembla. The men decided to take matters into their own hands. Using a hand-cranked drill, they began to seed cracks in the wall that lay before them, trying to wedge small gaps into fissures to break through the ice. Sections of the iceberg began to crumble. They moved the ships into the gap and pressed the ice ahead of them. In time sailors came to open water again, with only small, occasional floes to guard against. They cautiously raised their topsails and sailed on.

The sighting marked the end of the smooth sailing. From then on, ice could appear at any moment. The ships began a familiar jagged route of tack and tack again, as their course began to depend not only on the wind, but on the various ways the frozen sea could become impassable. Faced the next day with blocks of ice that filled the horizon before them, men braced themselves and surged again into the chasms between masses of ice, once more making their way through to the other side.

They didn't remember seeing this kind of ice this far west the summer before, and began to worry that the sea to the east had frozen solid, and that the ice might run all the way to shore. Based on the first expedition, they guessed they sat only two or three dozen miles out from the strait. But the passage between Vaigach and the mainland might already be closed for the winter.

On August 18, the ships came to the trio of islands at which they had reunited on the first voyage. Seeing familiar land was a tremendous relief, and boosted their confidence. They remembered that, on their way home during the last voyage, the stretch between the straits and the islands had sometimes been treacherous, but had at least been free of ice. They hoped to find navigation easier going forward. The sailors continued to vacillate between hope and anxiety right up until they came in sight of Vaigach Island and the strait. The sun was just above the horizon as they floated toward the entrance, giving them a clear look at their portal to the East.

Ice choked the whole span of the route, as if some pale extension of the mainland had overgrown the passage to form a new landscape. It was August 19. By the same date on their trip the prior year, van Linschoten and his companions had already made it through the strait, reunited with the company, and begun to head home. But on this trip, they'd lost weeks to bad weather and ice. And they still hadn't faced anything like what lay before them. Two thousand miles into the journey that had begun with such certainty of success, they lost heart.

If they couldn't sail through, Commander Nay decided, they would at least try to clear their way to the southern part of Vaigach Island and the Cape of the Idols, where they'd seen hundreds of carvings the year before. At the Cape, they could move out of the current, and decide what strategy to take. Perhaps they might find a way to go on.

It seemed impossible that they might have to turn back after having carried their cargo so far, but the strait was clogged with ice. Even if they somehow managed to get in, they might well get stuck. And if the ice froze them in farther along the western side, even if they turned around, they might not be able to make their way back.

The floes crowding on their way out of the strait rode the current toward the vast slabs that had already accumulated there, forcing some of the icebergs back on themselves in a ring. The sailors guided the fleet through the tightening whirlpool of ice and took refuge under the shelter of a point on the coast of Vaigach Island, where they dropped anchor for the night.

In the morning, a group of sailors went ashore from the convoy to investigate the terrain and get a fuller view of the ice blocking the waterways. They looked out to the west from where they'd just come, and saw ice everywhere, with no entrance other than the one they'd already made. They spotted a Russian ship near the shore. Meanwhile, the recognizable crack of gunfire traveled over the ice. The sound had come from the *Griffin*, a summons from the admiral to come aboard. They watched a group of men in the distance hurry onto the *ladya* and push off in alarm at the noise, leaving behind their fishing nets.

Going over to the beach, they found leather bags, whale oil, and a sled made of wooden sticks pegged together. Their orders from the States General were to engage only in good-faith trade, which meant leaving supplies like this as they found them. The crew hadn't spotted any homes or people near the straits so far on this voyage, but it seemed as if the area might be populated. Some ashore thought they spied the roofs of huts.

Meeting with the admiral later that day, the officers decided to go ashore and look for settlers and make a more thorough survey of the ice. They couldn't row across to the mainland because the strait to their south was too packed with ice even for their small yachts. But the next day, fifty-four men were dispatched with weapons to make a sweep of Vaigach Island. They spent the day ashore walking eight miles across the land and back without finding a living being. But they did come to a range of hills where they found more leather bags filled with whale oil half-hidden under rocks, along with reindeer bridles and walrus skins—the gray patches they'd mistaken for roofs the prior day.

Farther along, they discovered sleds piled high with the skins of reindeer and foxes, as well as tracks from not just animals but humans of every size and age. Their arrival must have interrupted a whole group of people, who were no doubt now hiding from the Dutchmen. The sailors realized they'd come to a trading post of sorts, where Russians visited with locals who had hunted and trapped and could barter with them for furs and whale oil. They set out bread and cheese, along with trinkets, near the items already on the ground. Finding no living creatures or signs of other outposts, they returned to the shore and made their way back to the ships.

Though the men had been instructed to leave things where they lay, the fleet's officers discovered that some of the crew had taken hides on board in violation of orders. Their crime went directly against those orders, though it was, perhaps, no worse than van Linschoten's theft of the wooden idol on the prior voyage, for which he was never punished. Common sailors, however, were entirely subject to any captain's wrath, doubly so the commander of a fleet. The perpetrators were sentenced to be keelhauled.

Keelhauling went at least as far back as the Greeks. The ordeal was sometimes administered as a punishment for piracy, the infraction committed at Vaigach Island. The Dutch practiced it more often than their contemporaries did, yet it remained brutal enough that it still wasn't common practice. The sailor in question was tied to a rope line and dragged underneath his ship from one side to the other. Significant injuries were to be expected. The penalty was especially brutal in Arctic weather with bitter temperatures in water and air alike, as well as the possibility of hitting ice during the journey under the ship. One of the sailors keelhauled off Vaigach Island didn't survive his punishment but was instead torn in half, another grim omen for a voyage that had already seen tragedy. The punishment would have serious repercussions, but the fleet went on with its mission.

One of the yachts soon returned from a day of exploring the strait, a feat made easier without the burden of keeping the larger ships of the convoy together. The boat had sailed as far as Cross Cape, where ice barred progress completely. The sailors then went ashore, where ice jammed even the waterway near the mainland. Far from shore, however, they reported seeing open water. They thought all the ships together might be able to make a run at clearing the strait. The men regretted not getting to talk to the reindeer herders, who might regularly cross the passage with the seasons and could advise them on what lay ahead.

Conditions in the bay where they sat at anchor grew more treacherous. The very point of the anchors was to limit the ships' mobility, but as a flotilla of icebergs began careening into the bay, anchors left them with no ability to dodge oncoming objects. They brought their ships closer to shore, which ended up trapping them even more. They began to think that they might have to drag the vessels aground in order to save them. In the meantime, the *Hope* sent a small boat to the island near the Cape of the Idols to get fresh water, but the water surrounding the boat froze over so quickly while they were ashore that the vessel got trapped. The sailors ended up abandoning half a dozen barrels of water, trying to fight their way back to the ship before they were entirely frozen in. The *Griffin*'s yacht had gone to the other side of the same island into the strait itself and was similarly besieged by ice. Unable to break through, they dragged their yacht ashore and left it, walking back over the island to the bay side on foot to rejoin the fleet.

That night, a fierce storm struck, driving the rest of the ships together and bringing the ice piling into the bay. But in the morning, they found that the tempest had loosened the small channel through the ice that had let them into the inlet. It now lay open. They dared to hope for similar conditions across the remainder of the strait. Maybe they could find a way to go on. Warm weather the next day farther encouraged them. On August 24, enough ice had melted to send a yacht on patrol again. They sailed back to the site where a Russian ship had been spotted before. Finding the sailors had returned to their ship, the Dutchmen saw that some were cooking barley flour on a fire while others skinned a walrus. They sailed closer, and the Russians came toward them without making any aggressive signs or threats.

The Russians described sailing from near the city of Archangel at the southern end of the White Sea. After spending all summer at the southern end of Nova Zembla trapped by the ice, they'd arrived on the strait only the day before. Another ship had been trapped with them farther north, and they were waiting to be reunited.

Seeking advice, the Dutch heard a litany covering most of what they'd been told the year before—that while conditions changed year to year, the ice would vanish for a period of ten weeks or so ahead of the true and brutal winter. The sea on the far side wouldn't be frozen, even in winter. They called the island Vaigach, pronouncing it "Way-gatz."

Vaigach lay south of Nova Zembla but was separate from it. Another strait existed between Nova Zembla and Vaigach, they explained, but it was impassable due to ice. Nobody lived year-round north of the mainland. Instead, everyone came during the summer to trade with the indigenous population from south of Vaigach, returning home before winter set in. These Russians had never seen the open sea on the other side of the strait, but said that their countrymen regularly took shipments to the far side of Vaigach Island and even beyond.

The carved figures they'd seen on both voyages were idols worshipped by the Nenets, the Russians claimed, whereas they were themselves Greek Orthodox Christians. The Russians didn't want to trade with the Dutchmen but accepted an old compass as a gift in thanks for their advice.

The sailors went back to the ships and waited for their other yacht to return from its second day of scouting. Near midnight, the report came that they'd seen more ice, but it was clearing, and before they'd turned to come back to the ship, they spotted a clear route for the convoy to sail.

The next morning, crews began to make their way east with high hopes. They expected some ice and weren't surprised when it rose again on their starboard side. Once into the sea, they sailed along the coast of Vaigach Island, thinking to head north of the shore along which the icebergs congregated.

In time, they cleared the strait altogether, and the coast of Vaigach began to arc away from them to the north. But as they sailed on, the ice visible on the starboard side of the ship began to swallow more and more of their view, coming closer and closer to the island coastline on their port side. The path of water ahead grew narrower and narrower until finally there was no route forward at all. Looking back, they sat in an expansive half-moon of ice curving from the shore of the island in front of them all the way to the mainland below. Everything that lay to the east was a plain of ice. They couldn't cut their way through the whole frozen sea. There was nowhere to go.

After a time, they turned from the sea back near the entrance of the strait they had driven their vessels through with so much hardship, only beginning to reckon with the failure of their quest. They anchored safely for the night, but in the morning, they saw the wall of ice advancing on them. They worked their way for miles along the shoreline, but the ice pursued them as they went. Within hours they had to set sail once more, returning to the bay that had sheltered them before. It was August 27 when they saw more ice riding the current toward them, coming to crowd them further. They retreated into perilous waters less than twenty feet deep to keep their hulls from being breached. And still the impassable wall moved closer. As if laying claim to the ships, ice began forming on the vessels themselves, locking the vessels inside a transparent skin a half-inch thick.

By the next day, the surface of the bay had turned solid. Men climbed down over the sides of the ships and found they could walk from the *Greyhound* to the *Hope* and the *Griffin* and find the going as solid as any street at home without fear of getting their feet wet. The window of weeks without ice promised by the Russians seemed to have vanished. The sailors commended themselves to God's mercy and waited. If weather didn't work its alchemy in the coming days, they might be stuck until spring. If that happened, they feared they would die.

On August 29, a Tuesday, they lay at anchor, biding their time. William Barents went to meet with the admiral at length. They had words over the fate of the expedition, with Barents rejecting the idea of turning for home and finding himself opposed by many of the other officers. He had tacked hundreds more miles in order to work his way around the ice on his previous expedition, giving up only when his men had refused to go on. They hadn't yet tried nearly as persistently to find a way into the sea beyond Vaigach Island. Barents returned to his ship with the matter still unsettled.

Nighttime brought rain and a storm, which they were desperate enough to welcome. They hoped it would melt some of the ice without sending any loose slabs to batter the ships. In the morning, they watched the weather clear and saw their narrow space had expanded again as the icebergs shifted away in the wake of the storm. The only likely exit—the way they'd come in—was still blocked. They remained trapped, but not so narrowly as before. The small circle of water they sat in had expanded.

William Barents boarded his yacht with the Amsterdam interpreter and a company of sailors and headed to Vaigach to investigate the view of the sea from higher land. Ashore, they caught sight of some twenty Nenets men, whose presence took them by surprise. Not expecting to find human company, they'd landed only nine men and were outnumbered. Their young interpreter stepped out unaccompanied and unarmed to meet the strangers. A lone Nenets man walked forward from the other group in turn. Just as he neared the interpreter, the Nenets representative took an arrow and notched it in his bow, as if to shoot his counterpart. The interpreter cried out in Russian, "Don't shoot! We are friends." In response, the archer dropped his bow to the ground.

The two greeted each other and knelt down, touching their heads to the earth, as the impromptu hosts bid the newcomers welcome. The conversation quickly turned to seeking advice on ice, navigating the strait, and what lay to the east. Pointing to the northeast, the Nenets responded that some five days' sail from there, they'd come to a great open sea, one that he knew himself, having sailed it as a captain with his king and his people.

Barents took these locals for people of reasonable judgment. Gerrit de Veer recorded in his journal of the voyage that the women and men dressed alike, had long hair, and wore reindeer hides with the fur against the skin. Bowlegged, flat-faced, and fleet-footed, they remained wary of the Dutchmen. Barents went back to the ship to make a report, and the next day, they returned to get more information. The sailors tried a second time to get a look at one of the Nenets bows, but were again refused. They brought ship's biscuit to the leader of the Nenets, who ate it and thanked them. The Nenets responded by taking the Dutchmen on sleds pulled by reindeer, which seemed faster than horses. Conversation stayed civil until one of the Dutchmen gave a demonstration of his musket, firing a shot out over the sea. The noise launched pandemonium until the Nenets could be reassured it wasn't meant as a threat. The visitors explained how their guns worked and demonstrated shooting a small stone two or three inches across from a distance, which the Nenets wondered over.

They bowed heads toward each other and left again, saluting their new acquaintances with a trumpet while the Nenets went off in their sleds. A few minutes later, one returned alone in a sled and rode up to the shore. He approached the small boat and went aboard to find a carved idol that one of the Dutchmen had carried onto the ship. With hundreds of the rough-carved images around, the Dutchmen hadn't taken any individual artifact to be of any particular value. Their host chided them for stealing something that his people revered, and they returned it to him.

The sailors shoreside headed back to the fleet. Admiral Nay again came to speak with the men on the *Hope,* to let them know that they had surveyed the strait for miles. It was choked with more ice than they'd seen yet, with icebergs running aground in waters they knew to be more than forty feet deep. Pushing floes and floating slabs out of the way was dangerous enough, but moving large icebergs that had settled on the bottom would be impossible.

Admiral Nay also spoke with Barents. The captain asked once more what Barents thought best. Barents again said he wanted the expedition to press on. The admiral wasn't convinced, and remained cautious about the idea of violating the instructions on their commission by splitting the fleet. He was also aware of the toll the ice had already taken on the sailors. Barents was the fleet's senior navigator, but Nay was the commander, and he had grim advice: "William Barents, mind what you say."[2]

More information came on September 1 from a talk between the Nenets and the interpreter for the *Griffin*, whose long residence in Russia made for more fluent conversation. Since the Nenets didn't spend the winter on the strait, they couldn't answer some of the most important questions put to them, but they offered that the deep seas on each side of the strait didn't freeze over in winter, though the area in and around the strait, once fully frozen, would stay solid until May.

The crews woke to clear weather on September 2, and weighed anchor. The *Hope* set sail with only minor tribulations, but the *Griffin* had a harder time, and the *Hope*'s yacht was so wedged in the ice that it had to be dragged out by kedging.

Like hoisting a mast at sea, kedging was an almost miraculous process and well known to any seasoned sailor. The kedge anchor was attached to a line, put on a small boat, and carefully rowed out some distance from the ship. If the weight of the anchor didn't tip the boat and drown anyone on the way out, the anchor was dropped, and sailors hauled on the line, not to bring the anchor toward the ship but to move the ship toward the anchor. The yacht's crew spent the day dragging themselves along as best they could and then carrying the kedge anchor out again.

They eventually moved the ship, but the yacht lost the anchor that had helped trap it in the ice and bent a second one. The wind then began to come in hard with a storm behind it. The rest of the fleet could go no farther and dropped anchor to ride out the weather and wait for the *Griffin* and the yacht from the *Hope*.

By the morning of September 3, the storm had stopped. The ice blocking the entrance where they had sailed in days before seemed to have cleared. The convoy set out as a group again and made it through the opening. They saw whales spouting water and felt sure they were close to an open sea that would carry them directly to China, if only they could get past the ice.

But then it grew hazy, from the water in front of them all the way to the horizon, and they lost the ability to tell one from the other. The weather remained warm, and they couldn't help but think the temperatures might reduce the threat from ice. Soon, however, they began to see the tallest icebergs they'd ever encountered. Then haze turned to a dark fog on the water, erasing visibility as it descended. Looking out from the deck, they couldn't even see the sailors on the next ship, or how close one prow lay to another's stern. The only visibility they had left was an occasional clearing far up in the sky, where they could sometimes spy the masts and topsails of other vessels and the high silhouettes of icebergs above them.

Without wind, they couldn't steer in the current and began feeling the impact from ice floes "which seemed to be mountains of steel and rocks of stone." They were thrown blindly into disarray, turned in different directions, unable to see one another, and strung out in tunnels of ice. As twilight came on, they called out from the deck and could sometimes hear other ships respond in the darkness, but they couldn't locate each other at all.

The fog lifted, and they rushed to find each other before night fully set in. Three ships reunited, still on the eastern side of the strait. Faraway shots rang out soon after, and the three found their way back to the rest of the convoy, which had managed to gather. To their relief, they caught sight of States Island in the distance, where last year's expedition had anchored and men had gone hunting for crystals ashore.

They made a beeline for familiar terrain and dropped anchor just as the sky unleashed a tempest. A mountain of ice rammed the

Hope, making the whole ship shudder as the men stared up at the iceberg in fear. Sailors spent half the night working with a grappling anchor to drag the ship off its frozen slopes. The other ships passed the night in much the same way, fending off or recovering from assaults by icebergs.

The morning of September 4, they looked out at the eastern sea they'd come to, after so much effort, in hopes of crossing, and saw, from their northwest across to their southeast, nothing but sheets of ice. They were nearly socked into their small harbor, with only a small opening to a slip of water between States Island and the mainland in which to sail. They moved into the opening, and navigated to that side of the island, where they made the ships fast to shore with a cable, to get away from the encroaching ice.

Those who'd been there before knew that the island had good hunting. Some men went ashore in search of Arctic hares, while the admiral called a meeting of the officers and merchants' representatives to assess their situation. The council agreed that the next day they'd try the plan Barents had advocated, one last attempt to fulfill their mission and cross the sea on whose shores they now sat. Come morning, they'd sail toward the ice, and look for a path through. But if they didn't find one, it would be God's will for them to return home. In case they were separated from one another's line of vision by ice or fog again, they set up a series of signs and signals by which they could track each other. They settled in for the night and prepared to go out and meet the worst the day could bring.

There was no need to sail out in the morning. The worst had arrived on their doorstep by daybreak. Before dawn, ice had filled their old harbor as well as the route by which they'd rounded the island, and was now moving in toward the shore. They shifted the fleet once more, now to the far end of the island, to try to distance themselves from the grinding advance. Tying the ships together "board to board" between cliffs and near land, they went ashore to look for any way out of the trap.

When they looked out from land onto the sea, the view was no better. The ice had pressed in so thick and without mercy that they couldn't see any water at all. Filled with despair, the sailors began to grumble, complaining that they'd never be free, and that those in charge had willfully risked the lives of everyone aboard. They were terrified of being stranded for the winter and couldn't imagine how they might survive.

Along with the rabbits they caught, the men spotted a white bear. Chased with guns, the creature left the island and climbed onto the ice, headed out to sea. The men wanted to follow but didn't dare do so. To fill the time during which there was nothing to do but wait, they began looking for crystals again, which some had seen on the prior year's visit to the islands.

On September 6, the weather continued to clear and warm a little, but there was no chance of leaving yet. The men took out the small boats again and sailed ashore at their whim, wandering the coast to occupy themselves. Those who wanted to gather more crystals headed not far from shore, kneeling and lying on the ground to dig, with dozens of other sailors and officers wandering nearby.

A cry rang out. The kneeling men looked up to see a sailor from the *Griffin*'s yacht rise up in the air, his neck clamped in the jaws of a polar bear. The huge creature had somehow approached the company unnoticed. As the sailor called for help and pulled out his knife, the mate who'd been digging beside him ran in fear. The bear bit away its prey's jaw and cheek while the bleeding man stabbed ineffectively at it. A group of sailors began to assemble, planning to run the bear off in time to save their shipmate.

The men ashore gathered themselves and ran at the bear with the weapons they had on hand, in the hopes of saving their mate's life, however unlikely salvation might be by then. Showing no signs of fear, the creature dropped her prize and ran directly at the group, driving them to flee in terror. The bear seized the slowest of those making their retreat—the boatswain from the *Hope*'s yacht— and tore him to pieces.

The fleeing men gathered on the shore. Their shouts and screams drew attention from the ships, which were close to land. More men rowed over in the small boats to help. William Barents arrived with the newcomers to make a group of thirty in all. The new arrivals pressed the men to band together with their weapons and kill the creature at once, while those who'd already seen two shipmates slaughtered advised getting additional weapons and taking the time to plot a more strategic approach.

Meanwhile, the bear returned to eating one of the sailors it had killed. The skipper of the yacht, its helmsman, and the purser stepped to the front of the group of men. Three times the helmsman and the skipper took aim and shot at the bear. Three times they missed, the accuracy and range of their guns leaving much to be desired. The purser set up for another try and shot the bear between the eyes. The animal didn't drop the man it held in its mouth, but the blow staggered it. The purser and a Scotsman drew their cutlasses and ran hard at the animal, breaking the blades of their swords as they struck. Still she kept hold of her prey. Struck again with a blade on the snout by the helmsman, she howled and went to the ground, where he leaped on her and slit her throat. Opening her stomach, they found only the pieces of the dead men inside, and assumed she'd been without food for some time.

The next day they buried the men in a shallow grave on States Island and skinned the bear. No one felt eager to go ashore, and those who did carried heavy arms and kept watch. The wind and ice pinned the ships in their makeshift harbor the next day, with an inch of ice forming over the small patches of water that remained.

On September 8 the admiral again called a meeting of the captains and navigators on the expedition. There was some discussion of pressing on. The officers on the *Hope* were of the opinion that the expedition had done as much as it could, and came out against the idea. The men from Amsterdam, however, wanted to leave two ships on the eastern side of the strait, where they'd winter and could explore the northern Nova Zemblan route, heading into the sea early the next season.

The admiral replied that the impromptu plan wasn't part of the orders they'd been given, and that if they wished to stay, they'd have to do it without his authority and find out what would happen if they made that choice. He refused to give his blessing under any circumstances. Despite the admiral's warning, Barents and his fellow officers from Amsterdam pushed the idea of wintering once more. The admiral thought they should either proceed or go home as a group, but didn't yet demand they turn for home.

The uncertainty was too much under the constant strain of terror in the ice. In just over a month, the crew had seen four shipmates drowned in the collision between ships, two eaten by a polar bear, and another killed by the admiral's chosen punishment. The crew had had enough. Faced with the possibilities of going on or being forced to stay the winter near States Island, they mutinied.

No record exists of whether the men brandished weapons, or took a hostage, or tried to seize a ship. They must have felt extraordinarily desperate, because even if they gained control of part or all of the fleet, there was likely nowhere for them to go. They were surrounded by ice.

The rebellion was put down the same day. After naming five men as ringleaders, Admiral Nay ordered them taken ashore and hanged on States Island.

Any ship can reliably be counted on to have a ready supply of rope. The men died in the place where the bear had eaten their friends, a place from which it wasn't at all clear that any of them could escape. The mass execution brought the death toll so far for the expedition to twelve.

Like keelhauling, mutiny had been around since ancient times. Julius Caesar personally put down a mutiny in 49 BCE, executing the leaders of long-serving forces who'd rebelled.[3] An order from a captain at sea was the law; even civilians were required to obey it.

The uncertainties of long voyages made possible by advances in navigation offered more time and space for anxieties to take root. On Easter in 1520, during Ferdinand Magellan's circumnavigation of the globe, captains in his fleet led a mutiny. It took time and trickery to regain the ships—he cut one of the anchor cables, setting one ship adrift—but in the end Magellan drew and quartered most of the mutineers, leaving one captain and a priest alive but marooned together on an island.

Nearly a century later, as Henry Hudson tried to find a northwest passage over North America, his crew mutinied more successfully, setting him and his teenage son adrift in a small boat with some provisions. Neither father nor son was seen again.

A mutiny was also a vote against the leadership of an expedition—a vote a sailor staked with his life. It was either such a regular event or such a shameful one that neither van Linschoten nor Gerrit de Veer mentioned the uprising or the hangings in their public accounts of the expedition. They likewise never mentioned the keelhauling for theft. Even Julius Caesar neglected to mention mutinies against himself.

Sailors might head out on a voyage without being told of the exact dangers that awaited them. Magellan misled his crew about the ambition of his expedition so as not to spook them. Columbus gave his crew a false tally of miles traveled each day, so they wouldn't realize how far from home they were.

In misery and weariness, the Dutch fleet tried to set sail on September 9, but with the wind against them, they couldn't get through the ice. The *Griffin* turned back to the island once more, keeping away from the ice and near the coastline, only to slam blindly into an underwater cliff. The crew of the ship from Rotterdam, thinking the *Griffin* was only stuck in the ice, came close to help and drove their own vessel onto the hidden rock shelf, too. As the ice moved in, the rest of the fleet was entirely occupied with sending barges and trying to tow them out of danger. Throwing ballast and even some of their merchandise overboard, they managed to lighten their loads, and dragged the two ships off the reef without damaging their hulls.

The next day, the fleet made its way around to the other side of the island, and on September 11, tried to escape its clutches altogether. They voted to sail all the way to the ice in the east one last time, to make sure there was no visible passage. They got as far as Cruyshoek, where the fleet had reunited before, but the wind pinned them there. Some of the men took a barge ashore onto Vaigach Island and found a decomposing dead whale with a sixteen-foot jawbone. They chopped it into pieces and carried part of it back to the ship, not only as a keepsake but also to prove the existence of an open sea on the eastern side of the strait.

Socked in at anchor for the night in a hard storm on the thirteenth, sailors watched the wind rage with enough fury to pull their barges and small rowboats off the decks and fling them on the ground. The next day they recovered all but a few oars that had been sitting loose in the open boats. At midday, they readied themselves to sail. On the fifteenth, the officers met with the admiral aboard the *Griffin*, with everyone acknowledging that navigation couldn't defeat or elude the threat all around them. The captains, navigators, and merchants' representatives signed a statement drafted by van Linschoten acknowledging that "because the Lord God has not wanted to allow for this journey, they find themselves forced (indeed and enough against their will, because of the lapse of time) to have to give up the same journey this time, due to the hindrance of the ice."

Once they began their retreat in earnest, the ice soon became less of an obstacle, but the approaching winter made itself felt in other ways. Their first day sailing for home began brightly, but devolved into snow, hail, and such ferocious winds that at times they could carry no more than a mainsail without fearing damage to the vessels. Snow and hail lacquered the vessels white, turning them into ghost ships.

By morning, each ship of the fleet found itself alone on the sea; they'd been scattered overnight in the storm. The *Hope*, with van Linschoten aboard, sailed on alone, and was eventually rejoined by its yacht. Barents in the *Greyhound* reunited with the *Griffin*'s yacht a day later, but they were soon separated again.

Barents spent two weeks fighting unhelpful winds making his way to Kildin Island, where conditions wouldn't let him get to land. The temperatures continued to drop each day and more hail fell as he continued sailing north. He arrived at Wardhuys, just shy of the Norwegian border, on September 30.

Van Linschoten in the *Hope* would come to Kildin days after Barents reached the same spot. But the crew of the *Hope* found it equally unattainable. In the end, they sailed past Kildin and Wardhuys as well, continuing on without stopping. Despite all the hail and storms, they nearly lost the ship to fire when a serving boy accidentally set the kitchen ablaze.

Darkness had swallowed more and more of each day, with only six hours of daylight left to them as they rounded Nordkinn. On the few clear nights, a captivating aurora borealis played in the vault of the sky. (If one had never heard about it before seeing it, wrote van Linschoten, "it would give one enough to think about.")

They'd been out nearly four months, and the men had begun to suffer stiffness in their legs and backs, as well as loose teeth and diseased gums. The officers recognized the signs of scurvy, but misattributed its cause as coming from the damp, cold conditions. The crew likely had little in the way of clothes, and certainly not enough in the way of winter garments for the Arctic voyage.

As they sailed closer and closer to home, the days grew slightly longer. The snow thinned then vanished altogether, with the occasional summer day taking them by surprise. Ships of the fleet staggered in separately between late October and mid-November, with no news about the other vessels along the way. It was impossible to know who might be delayed or lost until a ship glided into harbor or simply never returned. Barents, who hadn't wanted to come back that winter at all, was the last to appear, sailing into harbor on November 18, marking the safe arrival of all the ships.

They praised God for their survival, but noted that, for unknown reasons, their deity didn't want them to succeed on this trip, sending a long winter and excessive ice to end their journey. Van Linschoten argued that another trip during a milder summer could easily meet with success. The more difficult a thing is to accomplish, he noted, the greater the glory would be in achieving it. After all, the Portuguese didn't establish their Eastern trade empire on the first, second, or third voyage. They'd spent time and money investing in the project with a long view toward what success would mean, and as a consequence, were reaping untold riches from trade there.

The trick, van Linschoten argued, would be to find the right time of year in which to make the passage between Vaigach Island and the mainland, and to hit it perfectly. He advised sending two yachts to survey the area and gather intelligence on the weather and the currents before sending another convoy. But William Barents, who likewise advocated for a third voyage, would never return to Vaigach Island.

CHAPTER FOUR

Sailing for the Pole

When the Arctic fleet returned to the Netherlands for the second time, the jawbones of the dead whale were given to the cities of Enkhuizen and Haarlem as keepsakes. But no animal, living or dead, would have satisfied the burghers of the Netherlands, who hadn't expected to see the ships for at least a year—and had hoped they'd return carrying Chinese cargo. Instead, the investors wound up with two damaged ships, missing merchandise thrown overboard to get off a rock ledge, and the expenses of outfitting a seven-ship fleet with no return on the investment.

Despite the poor results from the second voyage, Barents immediately began proposing a third attempt. He prepared a prospectus for sailing to the eastern side of Nova Zembla, which he submitted to the Dutch provincial council. It was rejected.[1]

The other Dutch expedition sent to sail south around Africa when Barents had headed north for the second time hadn't returned yet. It remained to be seen whether the southern quest would succeed or fail. Business and political interests agreed that a northeast passage was still a worthy goal, but merchants had no intention of losing money. Instead, they offered a twenty-five-thousand-florin prize (worth nearly half a million US dollars in twenty-first-century currency). Potential explorers were also offered a share of the sale of imports for the first vessel to successfully sail a northern route to China.

Petrus Plancius, who'd created maps for both the northern and southern Dutch expeditions, also supported a third northern expedition, but this time recommended an even more radical course. It was clear that masses of ice sometimes congregated close to shore. Vaigach, near the mainland, had been choked with it. Barents had run into it in northern Nova Zembla. But the key might lie in finding a path through the ring of ice believed to surround the North Pole. Plancius thought the next expedition should set out from the Norwegian coast and cross over the top of the world.[2]

After the spectacular failures of the second voyage, Enkhuizen, Rotterdam, and Zeeland were unwilling to continue as patrons. But the city of Amsterdam agreed to make a third attempt. Two ships were chartered. Owing to the unrest on the first voyage and the open mutiny and executions on the second, unmarried men were sought for the trip. Warned that they might be at sea a long time, they were promised one wage for joining the voyage, and a second, larger purse if the expedition arrived in China. Without van Linschoten and the other cities' merchants to contend with, Plancius could effectively direct the route of the expedition toward the North Pole.

Jan Cornelis Rijp, who had sailed on the second voyage as a merchant's representative, was named captain of one ship. Jacob van Heemskerck, who had sailed in the same role for the representatives of the merchants of Amsterdam, was named captain of the second ship. William Barents would serve as navigator, sailing with van Heemskerck. Plancius could imagine how they *might* sail toward the Pole, but it would be Barents's responsibility to find a navigable route in the real world.

Goods would be sent for trade, in case van Heemskerck and Barents could deliver them to China. But this time, merchants wouldn't send their most valuable items, the things they'd most regret losing. They'd send second-tier wares instead. And the men would sail due north.

William Barents's ship on his third outing has no name in the historical record but was of a typical Dutch style known as a yacht. It was constructed of wooden planks wrapped and sealed around ribs and braced by knees. A cargo area sat at the bottom of the hull, which the crew crammed with a wide assortment of goods—mostly second-rate items, but a few impressive articles with which to initiate trade relations. Tucked among the rope and lumber that might be required for various repairs stood chests with cloth, pewter housewares, and thousands of paper prints. Along with the cargo and spare parts, a small boat or two would be stored in pieces for use in the event of damage to or loss of the one on deck.

Fifteen men would ship out under Barents and van Heemskerck. If they weren't on duty or eating, they would sleep in rotations on the orlop deck above the cargo area. Crew members likely did so without mattresses, using their clothes for padding. They would've bedded down in makeshift bunks under and between portholes built to accommodate the wheeled cannons they carried in the event of an attack.

A ship at sea is a crowded space. The cook's stove, which sat on the same level where the men slept, was a low rectangular box that could be hoisted aloft for use on deck because of the smoke it generated in any confined space. But in the deep cold, sometimes the heat provided by the stove in the men's quarters could offer enough comfort to offset the irritation from the smoke. A hatch leading to the main deck offered the simplest way to load cargo. At the far end of the orlop, another hatch offered a set of stairs as another way to climb to the main deck. Topside, lattice work and beams added on to the main structure of the boat provided some additional protection from the elements. Slightly shielded from view, a section of open toilets perched near the bow, with captain's quarters inside a cabin at the stern of the ship.

With shipbuilding still an artisanal occupation in Barents's era, no plans of the vessel exist. But from illustrations made in the era, we know its general dimensions. A typical yacht, it stretched about sixty feet from bow to stern, and spanned sixteen feet at its broadest point. Its foremast and mainmast rose into the air sixty feet or more, with the mainmast the taller of the two. Crew members furled or unfurled sails by loosening ties that held them, hauling on lines to pull the cloth taut while leaving enough play to catch the wind. If every sail were unfurled at once, the ship would fly a half-dozen sheets of canvas, like a bird flashing all its plumage.

The two-ship fleet took less time to prepare than the large convoy the year before. Crews were complete for both ships by May 5; five days later they set sail. They'd wound their way out to the barrier islands of the Vlie by the thirteenth. With wind and tide against them, however, they couldn't escape. In one attempt to set out, Rijp's ship went aground. Once he managed to leverage the craft off the bar, both ships anchored at the Vlie, where they were pinned down for five more days.

Given a third chance, Barents had to know that the burghers of Amsterdam, as well as his own mortality, were unlikely to offer many more opportunities for Arctic seafaring. Many years older than van Heemskerck, he knew that his desire to find a northern passage was now pitted against the promise of the southern route. The southern course might be treacherous or incapable of being seized by the Dutch in that moment, but it had already been proven to exist.

On May 18, Barents finally sailed due north from the Netherlands. Both captains directed their ships toward the pole star for two weeks, and by June 1, they found themselves in the land of the midnight sun. They'd expected as much, but were more surprised three days later when they looked up from the deck into the sky and saw not one sun as usual, but three—"a wonderful phenomenon in the heavens." On each side of the large, regular sun sat another, smaller one. Two rainbows arched through the trio of discs, a third encircled them all, and a fourth one cut through the middle of the third.

The parhelia they witnessed were caused by the refraction of light on plate-like ice crystals in the atmosphere. In the Arctic, low-lying clouds of ice crystals known as "diamond dust" can sit suspended for days, triggering visions of false suns, or sun dogs, that mimic the real sun, as well as the rainbow halo connecting them that was seen by Barents's men.

Sailors were superstitious creatures, and the Dutch had their fair share of signs and omens. Ships that tilted slightly to starboard fully loaded were lucky. Water sprites could curse those who killed them. A mermaid was kept on display in the Royal Museum at the

Hague. It was later investigated in the eighteenth century by one minister, Mr. Valentyn, who reported that there could be no doubt about its authenticity. "If, after all this, there shall be found those who disbelieve the existence of such creatures as mermaids, let them please themselves."

As more ships headed south from Europe and crossed the equator, complex hazing rituals of dousing and fines for those who were making the journey for the first time would be invented. To honor Neptune, god of the sea, Dutch sailors fined any shipmate crossing the equator for the first time. If he couldn't pay the fine, the crew tied a rope around his torso, strung him up from the yardarm of the ship, and dunked him three times in the sea.[3]

Moving from a minor trading nation to a global maritime powerhouse in just a handful of years, the Netherlands produced sailors who'd craft many rituals of protection for unknown waters in foreign lands, but in the time of Barents's voyage they had few special ceremonies. It may have been a measure of confidence in their leadership that the sighting of the triple suns didn't provoke deep unease. Or the sailors might have already heard of the parhelia that shone above them. Aristotle had described them in his *Meteorology*. Hearing reports from witnesses, Roman orator Cicero called for an investigation into them in place of outright disbelief. Even for those who might have felt a twinge of awe in the face of the wonders they witnessed in the sky, multiple suns and rainbows would surely be a good omen.

Not that the expedition's leadership was in an agreeable mood. Hardly away from Europe, Barents and Rijp had already begun to quarrel over their bearings. Rijp refused to keep with Barents's ship because he thought they needed to keep sailing north, while Barents insisted they needed to turn farther toward the east. Rijp insisted that if he did so, the ships would be forced toward Vaigach

Island, and he wasn't willing to risk going through the strait again. Barents insisted that they were so far west that they'd already abandoned the planned northern route. The two men argued bitterly, with Barents eventually shifting course to keep with Rijp.

They were still shy of the summer solstice, but the weather was turning colder. On June 5, a sailor standing on the foredeck cried out to shipmates below that he'd spotted white swans. As they climbed above decks to look for a bevy of birds floating across the water, they realized that the sailor was wrong. The swans were splinters from an iceberg that loomed somewhere out of sight. Around midnight, they moved through the wedge of disenchanted ice, like some magic spell undone. The inevitable had happened: ice had returned. They were back in the north.

Their plight made itself clearer in daylight, when the crews found themselves up against walls and mountains of ice. For four hours, they shifted sails, moving southwest and west to escape the mass before them, skirting its edges in an attempt to find a way around it. When they reached 74 degrees of latitude north of the equator on June 7, they found the water as green as grass, and moved between vast icebergs, as if land and sea had reversed. They thought perhaps they'd come to Greenland.

The ships approached a maze of ice too dense to be navigated. The on-deck hourglass ran in thirty-minute increments, turned again and again for hours as they sailed southwest, south-southwest, and finally directly south back the way they'd come to look for an opening. On June 9, an island that didn't exist on their map—or any map—rose up before them.

The next morning, eight sailors rowed their small boat ashore. As they passed by Rijp's ship, the same number of men rowed over from Rijp's vessel to Barents's ship. Barents asked Rijp's navigator whether he now agreed that the fleet was too far to the west, but

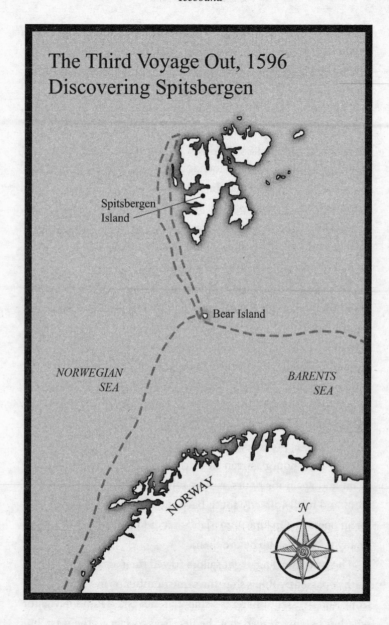

The Third Voyage Out, 1596
Discovering Spitsbergen

Spitsbergen
Island

Bear Island

*NORWEGIAN
SEA*

*BARENTS
SEA*

NORWAY

N

this question only provoked more ill will, spurring an argument. Barents swore that in time he'd prove that he was right.

Going to the island again the next day, the sailors found seagull eggs on the beach—a treat after weeks of eating food mostly stored in barrels. But even in the midst of such discoveries, every element of the voyage threatened to turn treacherous. Scaling a steep mountain of snow, a thing that didn't exist in the Netherlands, the island explorers turned around to realize their danger. Without experience climbing, they'd given no thought on the way up as to how they'd get down. The slope was too steep to allow them to return the way they'd gone up without falling. They couldn't devise a solution other than sliding down the mountain on their backsides without any control, despite their fear of breaking an arm or a leg on the jagged rocks at the foot of the slope. Standing aboard the ship and watching the men descend, William Barents was horrified.

They'd sailed once more into merciless terrain without even basic strategies to survive in it. Though they didn't have the skills that would've helped them most, they were growing accustomed to their deficiencies, and finding a way to press on. They also learned that danger could disappear as arbitrarily as it had arisen. After surviving the trip down the mountain, the sailors collected the eggs they'd found earlier in the day and gathered on Rijp's ship to eat them.

On June 12, they caught sight of a white bear in the water and set out in their rowboat to collar her. Drawing closer, they realized she was far too strong to subdue. Since abduction seemed impossible, they decided to kill her. Returning to the ship, they called for more men, along with muskets and arquebuses, the latter a long gun that had to be balanced on a stick to sight a target. Along with firearms, they brought hatchets and halberds—long poles with axe blades mounted on the end—and rowed their way back to the animal. One man swung an axe and buried it squarely in the bear's back. Yet the bear swam away. After chasing it in their two rowboats, they managed to corral the beast and split its head open. The battle lasted two hours, after which they skinned the twelve-foot pelt and tried eating some of the bear's flesh, which tasted unpleasant.

They weighed anchor and left the coast the next day, christening the place Beyren Eylandt—Bear Island. Heading north and, they thought, a whisper to the east, they lowered their weighted lead line more than six hundred feet, but the sea had no bottom. They spied a ship in the distance, but drawing closer, realized it was a dead whale. Covered in seagulls, the carcass reeked of decay.

Sailing into mist and gray weather, they could no longer tell sea, ice, and sky apart. Bits of floating ice released tiny bubbles of air, hissing like oil in a skillet. Larger blocks with air pockets growled at a lower register. Flat pancakes of ice moved with the tides, creaking like hinges. And vast icebergs ground against each other, whistling and keening through the friction like the winds of distant hurricanes. As they made their way north of Bear Island in fog or drizzle, the crew sometimes heard the ice before they could see it.

They sailed three more days, working their way around the ice, nearing one floating mountain so enormous they couldn't navigate around it, though they tacked and tacked again when the wind pushed them too close. On June 19, they spotted land. Taking the height of the sun, they worked out their location and found that they were more than a hundred miles north of the farthest northern point to which Barents had sailed on his first voyage, when he crested Nova Zembla—and likely well west of there. Barents was sure they were on the eastern coast of Greenland. What appeared before them was nothing like Bear Island—a tiny speck of land in the sea—but a substantial landmass that had never been recorded or mapped. They would later give it the name Spitsbergen—or pointed mountains—after the peaks they found lining its shores.

The coast stretched on and on for miles. They saw a good route to shore, but the wind held them at bay. Finally dropping anchor on June 21, both crews took their rowboats in to collect stones for ballast, without which the vessels would ride too high in the water and become unstable. Sometimes used to replace the weight of rations as they diminished or in trading ships as a placeholder for cargo until cargo was acquired, ballast could also be used to shift the balance of a listing vessel.

Barents's men were loading up their small boat when they spotted a polar bear on the water headed toward their ship. They ran to the men in the boat from Rijp's ship and climbed aboard, rowing out together to cut it off. The creature went farther into the sea, and they chased it, but it was faster. The sailors went mile after mile in pursuit, eventually using not only both small boats but an even smaller craft they thought might be nimbler. In the end, most of the crew from both ships had crowded into the boats and were hacking at the creature as they went by, drawing its blood and breaking their blades. The animal slashed Barents's small boat near the stern, landing a blow that sheared wood away and might have flipped it had it struck closer to the middle of the boat.

In time, they wore the bear down and killed it, dragging the corpse from the boat back to the ship. Though they had no taste for the meat, they skinned the beast and measured the pelt at thirteen feet. Slaughter emerged as the instinctive Dutch response to the Arctic landscape, a new theater that would see the same performance again and again with every European wave of arrivals. As historical archaeologist P. J. Capelotti observed about the killing of animals in the high Arctic that accompanied modern exploration, "It's amazing there's anything left alive."[4]

The crew took the smallest rowboat toward land and explored the coastline, heading into the mouth of a river with an island at its center. On the island, they discovered several barnacle geese sitting on nests. They killed one with a stone, and after others flew off, picked up some sixty eggs. Geese were far more appetizing than polar bear meat. They ate the bird they'd killed then carried the eggs back to the ship.

Along with being the first known humans to set foot on this newfound island, they'd accidentally solved a long-standing mystery about the geese. Because these birds vanished each year from their European habitats and returned the following year but were never seen laying or nesting their eggs, it had been a matter of folk superstition common in England and Europe since the twelfth century that they grew out of driftwood, or perhaps shells grown on a "barnacle tree" that fell in the water and matured. The theory had skeptics from the beginning but was accepted even by birding authorities well into Barents's time and beyond.

The records of the expedition directly discredit the theory: "It is not to bee wondered at that no man could tell where they lay their egges," Gerrit de Veer wrote, noting that as far as anyone knew, no one had ever traveled that far north and been able to see the nests for himself. Though the discovery was recorded on the voyage with an understanding of its importance, and Barents would be widely credited with upending the myth, centuries would pass before it died out entirely.

The Scientific Revolution was blossoming in Europe, promoting the idea that careful observation of the natural world would yield secrets about how it worked—secrets previously assumed by many authorities to be mystical in nature. In the years before Barents sailed, Copernicus had laid a path for understanding the unknown by collecting data to test and revise hypotheses about the heavens.

The first patent applications for a microscope and a telescope would both be received in 1608, a little over a decade later. Soon after that, Europe would begin to embrace the scientific method, with René Descartes believing even something as complicated as the mind could be understood through examining physical processes. Though Arab astronomer Hasan Ibn al-Haytham had posited the basics of the scientific method centuries before, Francis Bacon, advocating empirical observation and experimentation, laid the basis for a rational approach to the world that would come to shape nearly every branch of Western science in the seventeenth century. The modern conception of science was just taking root in Renaissance Europe.

As had been understood when the parhelion showed multiple suns in the sky, Barents and his crew recognized they could do more than wonder over the unknown parts of the landscape they'd entered. They took the height of the suns and rainbows they'd never encountered before, just as they did the heavenly bodies well known to them. They checked their instruments against one another. They took depth soundings of the sea. They recorded the length of polar bear pelts. They found walrus tusks along the coast and measured their weight as six pounds. Barents and his men went into the unknown not just as navigators and sailors but, also, as proto-scientists documenting a new world.

The day after their bird discovery, the crew returned to its more conventional mission, cranking the line around the horizontal winch that lay like an overturned spool at the back of the main deck to hoist the anchor again. But heading north, they quickly found their progress blocked by ice. After returning to the spot they'd just left, Barents worked his way along the western shore in search of other ways forward.

Some of the men went ashore to check their compass on land. Its usefulness lay in that it would point north and could orient a sailor even in a storm or darkness. But since it relied on magnetic attraction, the needle didn't point to true geographic north, but instead toward Earth's magnetic pole, which moved over time but always lay somewhere shy of true polar north.

When exactly this gap between magnetic and polar north was discovered isn't clear—in earlier maps the North Pole itself was generally depicted as a mountain of magnetite. But German compass- and clockmakers seem to have made accommodations for the discrepancy as early as the 1400s, and Christopher Columbus definitely understood it during his travels to North America. Barents knew the difference—the map made from the records of the voyage would still depict magnetic north as a magnetite mountain, but distinct and some distance away from the pole itself.

During the time of Barents's third voyage, magnetic north was located far to the west in a part of Arctic North America sailed and settled by the Inuit. More than two hundred years after Barents's death, the area would be mapped by British explorer William Parry, who got stuck in the ice while looking for a northwest passage to China and had to overwinter there.

After exploring part of what Barents thought might be Greenland, the sailors took a compass reading along the shore, noting a 16-degree variation from what might have been predicted, owing to the gap between Earth's magnetic and polar north. Meanwhile, the crew of the ship that hadn't yet anchored saw a polar bear swimming toward them. The animal moved as if to climb aboard, threatening an encounter the men knew by now could easily end with sailors overboard or dead. One man ran to grab a gun, but firing a sixteenth-century long gun wasn't a point-and-shoot process. It was an improvement over prior centuries, in which a lit match had to be held to a pan holding a small amount of gunpowder to start the fire that would pass through a touchhole and the main charge, leading to actual gunfire. But the weapon still took time to load, remained hard to aim, and sometimes didn't fire at all.

When the sailor finally did fire a shot, the bear, startled, made for the beach and the men who'd taken the compass reading. Those ashore had no weapons with them and would be easy prey. The crew steered the ship after the bear, while they made a racket to get the attention of their mates on land. Hearing their cries, the sailors on shore thought the ship had smashed onto the rocks. Everyone ended up frightened, including the bear, which went back into the water and swam off.

Pinned near shore by the wind the following day, both ships were unable to sail far enough north to clear the island they'd discovered, and wound up anchoring sixteen miles away on the other side of the inlet. The next day, June 25, they headed into the waterway, but it became apparent that their course wouldn't safely lead them to another sea. It took them two more days to tack their way out, repeatedly shifting the bow of the ship into the wind using sails to make some minimal progress back toward the sea.

On June 28, they worked their way around the far western point that had blocked their progress northward along the coast. By this point, they'd grown accustomed to wonders in the sky, but all at once, the air was full of birds. So many filled the sky that several crashed into their sails. Ice, however, remained the real danger. The ships made their way south and west to try to avoid the threat, with the effort eventually taking them far from land. After spending the next four days trying to find an open route north, they wound up back at Bear Island again. Rijp and his officers came over to Barents's ship to devise a plan.

The ice would clearly do them no favors. Rather than trying their luck pressing north into the unknown on an open sea that appeared to be blocked, Rijp wanted the ships to return to 80 degrees of latitude, hundreds of miles north, near where they'd gathered eggs from the nests of barnacle geese. Once there, they'd look along the coast for a waterway through to the island's eastern side.

As was so often the case, Barents disagreed about the best route. Given that they'd just come from the area Rijp wanted to sail back to, Barents thought the effort to return and find a river or strait that might take them inland to the other side of Spitsbergen would be futile. Instead, he wanted to revisit the promise of his first expedition by heading to Nova Zembla and sailing north of it into the eastern sea.

It was July 1, a week past the summer solstice. Four days remained until the first anniversary of their departure from the Netherlands the year before on the second Arctic voyage. On that expedition, Barents had accepted the fleet's decision to try to make a way south of Nova Zembla through the strait at Vaigach Island into the eastern sea. When that failed, he'd also accepted that he wouldn't be permitted to overwinter with one or two ships from the fleet, or to try his hand again at sailing north over Nova Zembla. He'd come so close on his first Arctic voyage. It must have seemed then as if only the bad luck of inclement weather and the reluctance of his men had held him back. He wanted to return to Nova Zembla.

Rijp and Barents had words for the last time and decided to go their separate ways. Rijp climbed back over the side of Barents's ship and down into his small boat. Soon the sound of oars carried him away to his ship and his men. Aboard his own ship, finally loosed from the obligation to accommodate anyone else's wishes about his route, William Barents prepared to set out with Captain Jacob van Heemskerck, fifteen other men, and a portion of the cargo. Rijp and Barents would never see each other again.

Barents immediately set a southward course away from the ice. The next day—July 2—he turned east toward Nova Zembla. He

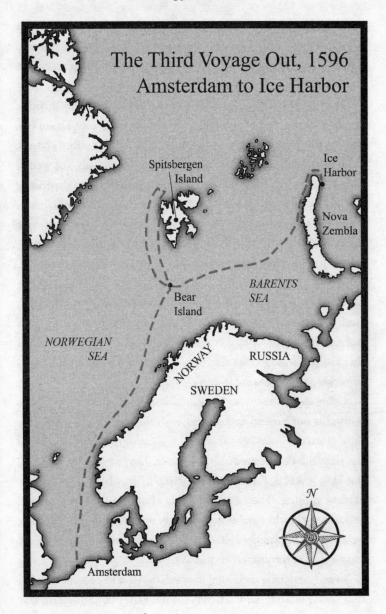

The Third Voyage Out, 1596
Amsterdam to Ice Harbor

Spitsbergen
Island

Ice
Harbor

Nova
Zembla

*BARENTS
SEA*

Bear
Island

*NORWEGIAN
SEA*

NORWAY

RUSSIA

SWEDEN

N

Amsterdam

wouldn't need to dip back down to the Norwegian coast and work his way across, as he had two years before. Knowing he was on a latitude that intersected with Nova Zembla, he could keep a course that was as close to due east as the ice permitted.

They were by now accustomed to weaving their way through frozen mountains, keeping their distance from the translucent palaces or monstrous shapes in the dusk, and shifting sails to retreat from unassailable walls that crowded in from the north. For weeks, they sailed more or less due east, edging south or north depending on the accommodation made by the ice, which went so far as to vanish from the horizon one morning, yet returned by sunset to drive them into a hasty retreat, as if under orders from an unseen enemy.

On July 14, they entered a canyon of ice, casting out their line, recording a depth of six hundred feet to the bottom, and then five hundred forty feet soon after. They made their way into the labyrinth without incident, but once inside, they couldn't find any path out except the way they'd entered. The breeze, which had helped them sail in, was against them once they turned back. It took hours of tacking into the wind just to escape.

Other days, they were becalmed and could make no progress with sails but instead had to drift with the ice. On the sixteenth, they spotted a polar bear floating on an iceberg. When it noticed the ship, the bear leaped into the water. They steered in its wake, thinking to kill it. But the animal made its way back onto the ice, eluding capture. They couldn't safely climb on the icebergs, and regardless, should anything go wrong, no one wanted to get left behind by a ship under sail. Out of disappointment or spite, they took a shot at the animal on the way by.

Even after losing their prey, they still thought its presence a good sign. Using floating ice as a barge, the bear, they thought, shouldn't be too far out at sea. They were surely nearing Nova Zembla again.

On July 17, Gerrit de Veer spotted land in the distance. Wanting to follow the coast, they turned to the northeast and raised nearly every sail to take advantage of the wind. The next day, they spied land again and recognized it as Admiralty Island. Two years earlier, Barents had passed it by because of its dangerous coastline. But he knew that it lay at almost the same latitude as Bear Island, which meant that it sat about two-thirds of the way up the western coast of Nova Zembla. Now confident of their bearings, the crew sailed on, soon coming to an island with two crosses upright in the ground—an island they'd also spotted on the first expedition. At that point, however, the ice asserted itself once more, and they couldn't work their way around it.

Dropping anchor, the men lowered their small boat into the water and rowed to land. The sailors visited one cross, where they stayed for a while before walking toward the other one. As they drew closer, two polar bears came into view. The Dutchmen were aware that the bears could smell farther than they could see, but they didn't know that the creatures could likely scent prey from a mile away, even buried in three feet of snow.

Soon enough, however, the sailors realized that the bears had noticed them. Rising on their hind legs, the animals moved toward van Heemskerck and his men. Failing to learn the lesson already offered more than once by the high Arctic, the crew had disembarked without a single gun. The sailors were in no mood to laugh at their predictable predicament. They made their way toward the small boat, watching to see if the bears kept on their trail. Their impulse was to run, but they'd seen bears outswim them in the water already. Even if the men made it to the boat, their chances on the water seemed slim. (The bears could in fact outswim the Dutchmen's small boats and, if necessary to survive, swim continuously for days.) On the last expedition, it had taken hours for three boats of men armed with guns to corner and kill one bear. A small boat on the water filled with men—many of whom were unlikely to be able to swim—could prove easy prey for two bears.

But the boat was all they had, if they could get to it. In one hand, van Heemskerck was carrying a boat hook, used for pushing vessels away or closer, or hoisting cables from the water. Addressing his men, he told them to stay together as a group and not run away. They should be safer clustered together and making a racket as a group to intimidate the creatures rather than trying to flee separately, which would make it easier for the bears to pick them off. The first man to bolt, he explained, would find the hook shoved between his ribs.

The captain and his men stole softly toward their boat, restraining the impulse to panic. They got to the craft and climbed in, having lost sight of the bears. Rowing back to the ship, they realized their good fortune. When they went back to the same spot the next day, armed to the teeth, there were no bears to be seen. But they found fresh footprints, showing that the bears had tracked them for a hundred paces or more. What would've been a massacre had the bears charged became a hair-raising story to tell their mates from the relative comfort of a ship edging its way north in the Arctic, or at home, if they were lucky enough to return. To leave a record of their voyage that would stand whatever the fate of the expedition, they built a third cross and made their marks on it. They had no shortage of time, because the ship remained stuck for almost two weeks. The sailors passed the days bleaching their dirty linens and taking care of menial tasks.

They didn't have to wonder for long where the bears had gone. Near the end of July, one came up to the ship, and they shot it in the foot to scare it away. The next day, another bear was spotted, and seven men combined their efforts to kill it and skin it, tossing the body into the water. On August 1, a third bear appeared in the vicinity but was spooked by the sailors' presence.

Three days later, they clawed their way out of the ice and made it to the other side of the island. Rowing a boat in, they began the grueling task of gathering more stones for ballast and getting them back to the ship. They set sail for Ice Point at the tip of Nova Zembla on August 5, with no ice in sight. As the wind shifted against them, they had to tack and tack again.

The weather turned misty the following evening, and visibility grew too poor to sail. The ship had come up near an iceberg that had wedged itself aground and seemed stuck. Over half of it lay invisible underwater, descending more than two hundred feet to the seafloor. The expanse visible above the waterline measured nearly a hundred feet more, dwarfing their wooden ship by comparison. With the fog already surrounding them, they were forced to moor the ship to the grounded iceberg.

They lay there the next day with fog haunting them. A snowstorm set in. As always, they kept watch. Late in the morning, Barents walked the deck in the driving snow and heard a snuffling sound. Looking over the side of the ship, he saw something climbing up. His cry of "A bear! A bear!" drew the crew from belowdeck. They made a huge commotion and frightened the bear, which turned and swam away.

Not long after, the creature returned, working its way behind the blocks of ice where they'd moored the ship. Visibility was poor, but the sailors kept watch as best they could. Pulling off the sail of their small boat, they used it as a tarpaulin to cover the deck. Four men crowded beneath it, holding guns trained in the direction of the ice. The bear suddenly appeared above them, having scaled the ice from behind. The animal moved to climb into the bow of the ship. The men fired a volley at the bear, shooting it in the body and

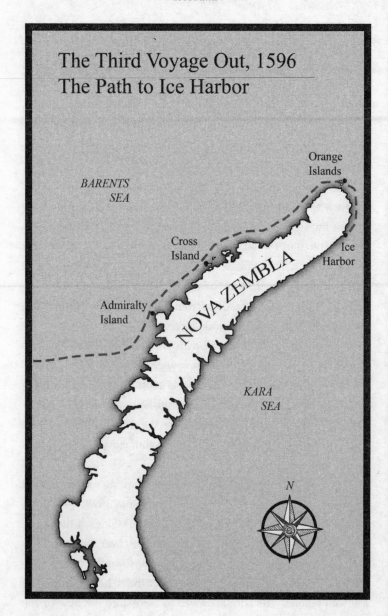

The Third Voyage Out, 1596
The Path to Ice Harbor

BARENTS
SEA

Orange
Islands

Cross
Island

Ice
Harbor

NOVA ZEMBLA

Admiralty
Island

KARA
SEA

N

driving it away. But they'd seen injured bears fight on before. The animal, they were sure, was hiding behind one of the many frozen slabs that crowned the field of ice before them. At times, they could hardly see in the storm, but they didn't spy any sign of the bear above them or in the water.

The next day, the ice field began to fracture, with currents shearing the frozen landscape. It was August 10. At one point, with the ice moving around them, they realized that the part of the iceberg to which they'd moored the ship still sat grounded, but most of the ice was in motion. Fearing they'd be crushed in the chaos, they unmoored the ship and set sail. Though the surface of the water was frozen, they skated over it, hearing the hull cracking through in places as they slid along.

Once in calmer waters, the crew used their kedge anchor to fasten the bow of the ship to another iceberg. When evening came, the new iceberg exploded without warning into hundreds of pieces that scattered far and wide. Slabs tumbled overhead, while whole submerged sections of ice burst underwater. Waves rose in the wake of the pieces that fell, and the Dutchmen worried that shards of ice might punch a hole in the hull or capsize their ship. Hauling their cable and anchor in, they escaped again.

The ship sailed a landscape full of wonders—but between the bears and the ice, no safe harbor existed. They made the vessel fast to a third iceberg and hoped for the best. Nearby, the spire of a frozen tower rose to a point more than seventy feet in the air, with a base that extended more than a hundred feet below the water.

On August 12, they had to move yet again to another iceberg, which they nicknamed Little Ice Point. They drew even closer to land to try to escape the drift ice, which ran with the current and whose bulk, like that of the stationary icebergs, often lay deep below the surface. If they kept to the shallows, the larger pieces of ice couldn't reach the ship.

Once close to land, however, they had to contend with bears again. While they fought the wind and couldn't move far easily, a bear came to the ship. They shot it in the leg. Fearing that it would return unannounced at a later date, they pursued it as it limped away, caught it, and killed it, adding its skin to their collection of pelts. The next day they finally broke free of the ice when the wind they'd been fighting eventually turned in their favor.

Sailing up and past the coast that fell away to the east, they crested Nova Zembla. By August 15, they were near the Orange Islands when ice came for them again. They managed to work the ship free and get to one of the islands, but when the wind changed directions, they had to shift the ship's position to keep it out of danger. Making a racket as they tried to save themselves from crashing into the rocks, the sailors woke a sleeping bear on shore.

The animal rose up and came for them. They had to abandon the work of turning the ship in order to fight the bear. But before they could kill it, they had to chase it into the water and onto the ice then back onto land again to catch it. After dispatching it, they returned to saving the ship. Whenever things looked bad, there was always something worse waiting to happen.

Once they were as safe as they could be while made fast to an iceberg that seemed to be lodged on land well below the waterline, William Barents sent ten men ashore in a boat. It was August 16. Climbing a high hill, they surveyed the view. With their backs to the North Pole, they saw only land straight ahead to the south. But to their left, toward the southeast, they saw open water. At the sight of a navigable sea, the sailors were sure they'd won the reward set out as a prize when they signed on for the voyage. They scrambled to return to the ship to tell Barents the news.

Hoping to make for that open sea, they prepared to sail in earnest on August 18. But ice broke in hard on them and would've swept them away if not for their kedge anchor and more than a thousand feet of cable they'd used to hold the ship in place. Barents could barely navigate the ship back to the place from which they'd been swept away. They decided to spend the night there, which still passed without any true darkness. The next day, they had good weather and set out again, turning slightly south along the coast, and spotted a bleak line of dark earth that stretched out into the water like a long, narrow finger. They named it Cape Desire.

Making steady progress, the crew had great expectations of reaching the open sea at any moment. They managed to work their way around the point of Cape Desire before the ice crowded them in once more, forcing them back toward shore. They crept their way south along the coast, which began to curve back toward the west as they descended in latitude.

They knew they'd entered the sea east of Nova Zembla, but the ice wouldn't let them break free to sail across it. On they crawled, close to land, naming as they went. On August 21, they were able to anchor at Ice Harbor, some fifty miles from where they'd rounded the northernmost part of the island. The next morning as they set out, the current drew them hard to the east. The wind began to blow. They moored themselves to a grounded iceberg once again—one that was entirely a shade of blue they'd never seen before. Climbing sixty feet above the waterline until they were on top of it, they found a layer of dirt and eggs. They waited out the storm, and an argument broke out over whether what they sat on was truly an iceberg or a shelf of land.

They headed out on the twenty-third, but the frozen yet mobile landscape pressed them back to Ice Harbor. Even in the harbor, the wind drove ice toward them, breaking both the tiller and rudder on their vessel. Pinned between their ship and the frozen wall crowding them in, their small boat was smashed. It seemed the ship itself might meet the same fate. They were stuck fast in place until August 25, when the weather grew calmer. The men decided to try to break up some of the ice and hack themselves free, but had no success. By mid-afternoon, however, the calved icebergs began to move on the current, without any help from the sailors. Though it didn't seem possible to go directly east, they could at least continue south along the coast and follow it down to Vaigach Island. If no opportunity to go in a more eastward direction made itself available, they could sail around the southern end of Nova Zembla and return home.

The plan was hardly hatched before it was spoiled. The wind was of no help, and the crew couldn't cut a path in the frozen sea to the south any more than it could to the east. The water had hardened everywhere against them, and any southward navigation became impossible. The next day, a gale rose up that seemed as if it could carry them north. Barents decided they'd give up on the southern course, and instead would try to get back around Cape Desire, making their way home by heading back the way they came.

As they approached Ice Harbor, however, their northward progress was checked. The wind past that point was a gale that drove the drift ice to turn solid all around them, even as they sailed. Three sailors climbed overboard and began hacking with hand-axes and a pickaxe to try to free the ship. As the ship surged back into the current, the men were nearly left behind. But the current pulled the ice they stood on in the same direction, and the crew members made a leap for the ship. One man held on to the beakhead around the bowsprit on the prow of the ship. Farther back, another caught hold of a cable strung through the corner of one of the sails. And a third grabbed hold of the end of a line that ran from the mainsail down to the stern of the ship. The sailors held on for dear life.

Despite the close call, all three were saved. They could control almost nothing about where the ice would take or release the ship. If the men hadn't caught hold at just that moment, the ship would've shot off, with no way to circle back to save them. They would've been left on the ice to die.

That evening, the ship finally reached the western side of Ice Harbor and dropped anchor. On August 27, the weather turned fair, but drift ice closed the ship in again. Sailors lowered one of the small boats and headed to land. As the wind picked up, icebergs drove into the harbor toward the ship, slipping under the bow of the ship. Lifting it four feet straight up, the iceberg began tipping the ship backward. The stern had been pushed down so far, it felt like it might be touching ground, or ice. The entire ship tilted unsteadily. Those still aboard put out a signal flag to catch the attention of the men ashore. Expecting they'd soon be submerged, the rest of the crew abandoned ship, fleeing in the other small boat.

After the crew had reassembled in the water, Barents got them back on board to inspect the ship. It hadn't, as they feared, overturned or been crushed. By the next day, the ice had relaxed its grip a little, allowing the boat to right itself somewhat. Barents and the other pilot went off the ship to inspect the bow, getting down on the ice on their knees and elbows to check its position and displacement. In the midst of their measurement, a loud crack sounded. The ship overhead sprang off the ice so suddenly that Barents and the pilot thought they'd be lost at sea.

But the ship, though it now sat upright in the water, was still trapped. Barents and the pilot made their way back aboard. The situation was no better a day later, and so on August 29, they began to work at the solid sea with crowbars. Making little headway, they commended themselves to the mercy of God. But the next day, the current from the sea they couldn't reach drove in more ice, grinding blocks and pieces against each other, packing them even closer together, and clamping the ship in a vise. A snowstorm set in, and under the pressure of the new ice and the wind chill, the ice began to crack and groan all around them. They could only watch and listen in terror as they sat aboard and felt the ship slowly rise out of the water.

By the time the storm was done and conditions had settled enough for them to explore their plight, Barents saw that, again, the front end of the ship had been lifted so that the bow sat several feet above where it would when floating on open water. Yet the stern of the ship hadn't risen with it, but instead appeared to be stuck in a crevice of ice. Rather than protecting the tiller and the rudder, getting wedged in the crevice had somehow helped to shear the steering mechanism off, breaking it into pieces again. But the part of the stern that remained wedged in the crevice appeared to keep the entire ship from being lifted up and run aground while yet more ice was driven into the harbor. They pulled out the small boats and set them on the ice away from the ship, in case it capsized.

Later in the day, the current pulled the new ice out. The ship eventually came back to level and sat upright in the water once more. But the men had no time to celebrate. They quickly set about making a new rudder and tiller, which they hung so that the equipment could be easily protected if the ship were wrenched upward once more. Water ran hundreds of feet deep not far away, but there was still no path for them on the surface of the sea. Their return to Nova Zembla had so far consisted largely of a series of dangers punctuated by temporary relief that didn't offer salvation but only a return to their former dire straits.

The next day was Sunday; they held a prayer service. Even as they prayed, the current drove the ice back in, pressing on the wood in and outside the ship. The vessel began to rise again. The men feared for the hull, but the hull held. The crew dragged the small boats once more onto the ice then up onto land to keep them safe and at the ready.

September 2 brought more of the same, as the ship was vaulted even higher by the ice. A blizzard began, and they considered using a small boat to carry barrels of bread and wine ashore, where at least they wouldn't have to worry about death by iceberg or drowning.

The next day, still sitting in high wind but with less snow, the ship began to pull from its mooring, and lost the cable the men had used to fasten it to the ice. The ice-knees that had been built onto the sternpost to protect the hull from collisions started to buckle, but their collapse was stopped by wooden planking that managed to keep them attached, though only barely. Despite the new ice surrounding the ship, the hull held and didn't flood belowdeck, which they took as a wonder, because of the mountains hundreds of feet high pressing in toward them.

Barents and the crew waited three days in fear as the weather slowly improved, with ice surging and withdrawing like the tide. After dinner on September 5, the ship began to list hard to one side. The hull held again, but there was no simple way to right the vessel on the seething ice more than a thousand miles from home. Any one of the cataclysmic iceberg explosions they'd already seen might drop it into water that could sink a listing ship. After a meeting to confer on their options, they took their old foresail and used it to transport powder, lead, small and large guns, and some furniture ashore. Making a tent of it under an overturned boat, they stocked it with bread and wine. They also brought tools to mend the small boat that had been crushed, as it might become particularly valuable if the ship were lost to them.

They spent most of the following week shuttling between land and the ship. Some of the crew went inland and found a fresh water source, as well as wood that had washed up on the shore. Meanwhile, the ice-knees protecting the stern finally gave way, and the ship lay encased in more than three feet of ice. Two bears approached the ship by night, but were scared off by the sound of the ship's trumpet and shots fired in their direction. There was no relief in sight.

On September 11, a party of eight men went ashore armed to the teeth. They verified that there was fresh water and found the wood that the scouts had described. The ship lay pinned as it had for two weeks, with the fall equinox coming on. Twilight would soon give way to night, and behind it, winter. Their spectacular efforts, which had saved them in every crisis so far—sometimes flailing, sometimes inspired, and often both—couldn't draw the ship from the ice or carve a path in the frozen sea. The sailors assembled together again and acknowledged that God wouldn't intervene to aid their departure. Gerrit de Veer wrote in his journal that Barents and his men began making plans to spend the winter on Nova Zembla "in great cold, poverty, misery, and griefe."

CHAPTER FIVE

Castaways

Hope of rescue is the sustaining dream of castaways. In May 1630, the commander of *The Salutation*, a British vessel in Arctic Greenland, sent out a party of eight in a small boat with "a musket, two lances, a tinderbox, and a brace of dogs." They were told to hunt deer as part of provisioning of the ship for its return voyage. While the men were on shore, ice moved in and drove the ship farther out to sea. Not knowing why the vessel had vanished, the men decided to make their way to where another ship in their fleet had been ordered to anchor for several days. Arriving to find that ship gone as well, they tried to get to a third ship that was supposed to leave with them and return to England—only to realize that without a compass or a pilot, they'd overshot the last reunion point altogether. Though they had only the clothes on their back and no food except what they could catch or find, they believed that help would be sent eventually. Despite tremendous suffering, they managed to survive for most of a year, until a rescue ship arrived.[1]

More than two centuries later, Sir John Franklin set sail from Britain in *The Terror* and *The Erebus*, both battle-ready bomb ships with hulls further reinforced to withstand collisions with icebergs. Franklin had already made his way to the Arctic on two earlier trips, the first of which aimed, as Barents had planned long before him, to go from the Atlantic to the Pacific by sailing near the North Pole. The plan worked no better for him than it did for Barents, but in 1845, he sailed over northern Canada in two ships with a crew of one hundred thirty-four men and at least three years of provisions in search of a northwest passage to the Far East. The expedition carried with it extraordinary technology of the day: steam engines to power the ship when the wind didn't cooperate and desalinization equipment that could transform the sea into drinkable water. At the end of their first year out, they wintered not far from Beechey Island in what would become Nunavut, Canada. With plenty of food and coal for fuel—as well as a central-heating system and a twelve-hundred-book library—they managed their first winter. But following the thaw in the spring of 1846, they sailed into Victoria Strait and were trapped by ice. They spent the next two years descending from disappointment into horror. No one aboard was ever heard from again.

Two rescue expeditions were sent out in 1848, one on land and one by sea, but they found no signs of Franklin or his party. In 1850, Franklin's wife sponsored a third expedition to search for the ships; several more attempts to find the lost sailors followed.

No news came until John Rae—who had led the first overland search party—returned to map the Canadian coastline in 1854 and heard a story from the local Inuit. They talked about men whose large ships remained trapped in the ice until their food had run out. Before the last of them had died, the Inuit said, the men had resorted to eating each other.

When Rae came home to England, he scandalized his countrymen by suggesting that a national saint had been reduced to cannibalism. Attacked over his account, he defended the indigenous Inuit he'd interviewed about the Franklin party, painting them as "a bright example to the most civilized people," as well as "honest and trustworthy," contrasting them with the desperate state to which Franklin's men would've been reduced as they sat trapped in the ice. When critics provided examples of British castaways who died without eating each other, Rae noted that typically in a shipwreck, the more immediate demon of thirst would displace that of hunger. But with the desalinization equipment Franklin's group had on hand, thirst wouldn't have been an issue. Fully hydrated, they would've had nothing but time to think about but their hunger, in a place that offered only "a barren waste with scarcely a blade of grass upon it."[2]

Rae's reports of cannibalism infuriated novelist Charles Dickens, who dissected them by detailing the near-death hardships that Franklin and his partners had endured on previous expeditions without eating human flesh. Dickens explained for his readers that "the lost Arctic voyagers were carefully selected for the service," dismissing the oral reports from the Inuit as "Esquimaux kettlestories." He listed secondhand accounts in which British explorers claimed to have held fast, while Native Americans had advocated cannibalism, implying that perhaps the "Esquimaux" themselves had eaten the crew.

At the time of its departure, all England had its eyes on the Franklin expedition. Once they were trapped and sent no news home, the castaways might have hoped that rescuers would find them eventually. This hope likely kept them going for a time, even as their first year frozen in turned into a second year then approached a third. But despite years of provisions and the beginnings of modern equipment for defying Arctic conditions, they couldn't hold out. Over a hundred fifty years after the expedition vanished, after more than a dozen rescue or recovery expeditions, casual observations by local Inuit would be the means by which both shipwrecks were discovered.

The need for rescue from a place that had a local population underlined how alien Europeans were to the Arctic terrain in which they became castaways. By the time the Franklin expedition vanished, Inuit were already living in the region, having migrated into the northern reaches of North America and Greenland centuries before. They'd learned how to fish and hunt whales, and traded with groups they encountered going all the way back to the Viking era during the Norsemen's westernmost voyages. But the Arctic was their home. It had dangers but didn't hold the same terrors for them that it did for European sailors in the post-Viking era, who typically wanted only to pass through the region unscathed.

In time, some explorers would come to recognize that Arctic populations would be helpful sources of training for polar expeditions. As he crossed the interior of Greenland in 1888, Norwegian explorer Fridtjof Nansen took a very different tack. He brought two Sami crewmen from northern Norway along with him. He carried Inuit-style snowshoes, adapted Lapp skis, and could speak Inuit. Most important, unlike the massive company assembled for the Franklin expedition, Nansen organized only small bands of collaborators for his trips and used indigenous survival tactics. In a testament to his approach, every man and piece of equipment he ever set out with came back.[3]

In the 1860s, American explorers Isaac Israel Hayes and Charles Francis Hall had headed on separate expeditions to Arctic Canada with the aid of Inuit men and women. Hall stayed five years before returning home. At the beginning of his second Greenland expedition in 1891, American explorer Robert Peary lived ashore for a time with a broken leg, communicating with the local Inughuit, the world's northernmost group of Inuit. Later, on his expeditions even farther north, Peary moved whole indigenous families to key transit points for support, having the women sew his clothing and the men accompany him on the final legs of each trek.[4]

Others would go without local help and stumble their way to death or glory. American Walter Wellman, a journalist with no Arctic experience, announced in February 1894 in the *Philadelphia Inquirer* that he would set out for the pole. Wellman had no interest in immersing himself long term in the Arctic and understanding its profundities. "What Wellman was truly after was a shortcut to the North Pole," wrote P. J. Capelotti.[5] Wellman imagined that superior technology, from aluminum sleds to hydrogen airships, could provide that shortcut. And though he wasn't above bringing along veterans of others' expeditions, he wasn't interested in picking up skills from indigenous people. His airship engine self-destructed when it was turned on in 1906, and the rebuilt models failed him twice in subsequent years, getting just a few miles north of the location at the northern end of Spitsbergen where William Barents had been forced to turn back more than three hundred years earlier.

Barents and his men didn't have Wellman's aluminum boats.

Nor did the Dutchmen have the steam engines, heated rooms, and library that Franklin had at his disposal. They'd observed the Nenets people they met—their use of sleds and their clothing—and sought their advice on navigation. But they didn't seem to have thought to emulate their ways in adapting to the climate. To a man, the Dutchmen stranded on Nova Zembla were phenomenally unprepared to overwinter in the Arctic.

They'd experienced cold winters in the Netherlands as mountain glaciers expanded in the Alps and elsewhere, causing a Little Ice Age to cool Europe. Yet they lacked real Arctic clothing; the men weren't provided any particular gear when they signed on to sail with the expedition. And even experienced sailors knew little to nothing about survival in the far north. They skinned polar bears, but apparently the hides were seen as so valuable as trophies or gifts for patrons that they never stitched the fur pelts into protection for themselves. Despite Barents's desire to overwinter on the second voyage, and how close ships came to getting frozen in on the prior trips, history records no special materials being brought along in the case of this eventuality.

Though its scenery might appear monotonous at first glance, daily life in the high Arctic involves constant change. A world of vegetation lies close to the ground—dwarf bushes, mosses, lichens, and yellow Arctic poppies, which, at three inches tall, tower like giants over much of the landscape. The sky is just as subtle. The clouds over one area of land often differ from those over an adjacent fjord, which stand distinct from the vapor moving over the open sea. This combination makes for volatile weather.

Equipment fails. The tide doesn't cooperate. Wind shifts. Things may not go wrong any more often in the Arctic than they do in other landscapes, but in the far north misfortune is far more likely to have a cascade of consequences. In the Arctic, a single wrong thing triggers several more wrong things, because food and shelter are so contingent.

Even knowledge of Arctic survival methods was no guarantee of success. An expedition that would set out for the North Pole in 1871 aboard the *Polaris* was led by Charles Francis Hall, who had lived with the Inuit. But the voyage quickly descended into crisis when a schism between the Germans and Americans aboard became open conflict, and Hall fell ill and died, after saying he'd been poisoned. More than a dozen members of the expedition were separated from the ship and drifted on an ice floe for months before they were rescued. The remaining crew and ship were likewise stranded after the vessel ran aground, forcing its passengers to overwinter and pray for deliverance, which came the next summer.[6]

But the sailors stranded on Nova Zembla had no hope of rescue. By the time Barents and his men made plans to spend the winter, Jan Cornelis Rijp and his crew were tacking somewhere north of the island they'd named Spitsbergen, headed for the North Pole or even China. Or maybe they were themselves stranded. Or perhaps they'd turned back before the ice could take them and had already made it home.

Wherever Rijp had ended up, no passing ship would sail by the shores of Nova Zembla at Ice Harbor, and no Sami or Nenets with reindeer would venture so far north. Even if Rijp tried to send help, no one would know where to find them. Their location appeared on no chart; they were off every map in existence.

On his first Arctic voyage, Barents had pressed eastward until his men refused to go farther. On his second trip, Barents had argued for staying behind with two ships to overwinter and scout out clear passage at the first spring thaw—a plan that likely helped provoke open mutiny and executions. On this, the third voyage, he'd finally sailed the route he'd hoped to with no one forcing him home, and now he would overwinter. As in some dark fairy tale, he received everything he'd asked for, but none of it came as good tidings. The issue of mutiny, which had haunted his prior voyages, was finally transcended—but only because any possibility of sailing for home had vanished.

Seventeen men remained. Along with William Barents as navigator for the expedition, Jacob van Heemskerck captained the ship and had control over the cargo. As he had on the second trip, Gerrit de Veer was present to record the events of the voyage. The rest of the crew included a pilot, who had some fraction of Barents's navigation skills, and a barber-surgeon—both with the last name Vos. Claes Andries was joined by his nephew, John. As was typical, the voyage had a cook, a carpenter, a gunner in charge of the cannons, and a variety of sailors, along with a ship's boy charged with helping to hoist and lower sails and maintain the ship.

The ship lay too vulnerable to the elements to protect the men for the winter. Even if the hull could've provided enough shelter, the ice threatened daily to destroy it, making the first order of business to build a cabin. Every sailor had basic woodworking skills, and even at sea, a ship's carpenter could nearly do magic. With the broad range of tools they had with them, the task of building a home would be a challenge but far from impossible. The bigger difficulty lay in finding enough wood along the coastline of an Arctic desert that couldn't grow anything taller than miniature shrubs. The entire trees they'd seen on prior voyages washed ashore from the continent were a promising start, but looking closer to where the ship lay, they found little else to help them.

Their first days on land, they were spared snow and rain, but fog sometimes forced them to stay aboard the ship or near the shore, peering into the mist and listening for the crunch of paws on ice. Even on a clear day, polar bears might smell the crew before the crew could see the bears. In a blanket of fog, the men would have no chance at all.

On September 14, sunshine made it possible to hike out to gather driftwood, which they began to drag into piles that could be easily found if snow covered the land again. Early the next morning, the cook had a tub of salted beef put out on the shore. Once a barrel of salted meat was opened, it had to be soaked in fresh water for hours to leach the salt from it and rehydrate it; then hours more of cooking were required to make it edible. Along with salted meat, their provisions included smoked bacon and ham, as well as fish, both dried and smoked. Every vessel carried ship's biscuit—the omnipresent food of European voyages, because it never went bad, being made of only flour and water and perhaps a little salt. The sailors also ate bread loaves, groats, barley, peas, and beans. They had oil and vinegar to cook with, salt and mustard to add flavor, and to drink (in addition to water) there was beer, wine, and brandy.

Though not intended as such, an open barrel of meat made good bait for bears, bringing three unwelcome creatures down to the shore. Spotting the visitors, the sailor on watch called his mates to come on deck and bring their guns. One bear lay at a distance, behind a piece of ice, while two more approached the ship. The closest animal bobbed its head into the tub and paid dearly for its ambition: a shot in the head killed it outright. The second bear sniffed the fallen creature and pondered it a moment before running off. The sailors watched for its return. When it came back at them, rearing onto its hind legs, they shot it in the belly, which sent it fleeing again. Taking up the carcass of the dead bear, the sailors gutted it and stood the animal upright to freeze. Though the surface of the sea had turned solid, more than an inch thick all around the blocks of ice, they still fantasized that they'd somehow get the ship loose and carry the frozen beast back to Amsterdam.

The sky remained in twilight rather than full darkness, but nights were growing colder. In the daytime, ice would thaw into water, but as the sun dimmed, the surface of the water froze again before morning. They realized they should start constructing a shelter as soon as possible. Some of the wood they'd found included large branches and trees, making one of the first orders of business the building of sleds to haul lumber nearer to the ship.

On September 17, they made their first trip with the sleds, dragging four logs for miles over ice and snow. When the sun rose again, more than a dozen crew members left with the sleds on another run, piling beams higher, with five men pulling each sled from the front and three staying behind to cut the wood down to logs that could be moved on the sleds. With more people, they could make two round trips each day, starting a lumberyard near the building site.

They worked wearing trousers that stopped at the calf, loose tunic-style shirts, as well as hats, jackets, and leather slippers—but no heavy winter coats. On September 18 they hauled wood in driving snow, and once the weather cleared, moved even more loads during the next two days.

While they worked on framing the cabin, sailors still spent the nights aboard the ship. The ship's stove, which was portable, was carried on deck to feed the men, but it grew so cold on September 21 that everything outside froze, and they had to bring it belowdeck.

There was little surprise in the temperatures dropping perceptibly: the fall equinox arrived the next day. The sun, which had stood so high, spinning a tight circle all summer in the dome of the sky, had been slowly making wider loops like a top winding down, swinging closer and closer to the horizon each evening then barely grazing it. That evening it vanished again just for a moment into darkness.

The return of night was ominous, but they'd expected it to happen. The surprise came the next day, on September 23. Weary from dragging sleds full of wood to build a shelter, several crew members trudged back over the rough snow and ice, only to find that the carpenter had died.

He'd come to Nova Zembla from the town of Purmerend bordering Lake Purmer, a few miles north of Amsterdam. He'd instantly become the most valuable person among the castaways in terms of assuring their survival, but he was the first to succumb to the brutal conditions. In a bad omen for the house they were to build, the ground was already frozen too solid to dig a grave. Short on time and with few options available, they tucked his body beneath the overhanging rock of a crevice on a hill the next morning, leaving his corpse to the elements and the bears before taking the sleds to get more wood.

Bitter cold had become the baseline, but the sky had at least remained clear for a few days when a fierce gale blew toward the west. On a flat section of peninsula, they plotted a rectangle thirty-six feet long and just over twenty-two feet wide. To make a level base, they gathered and spread flat stones. With the ground frozen so hard, the idea of using familiar Dutch construction methods or digging a foundation became as impossible as it had been to give their carpenter any proper burial. Instead, they likely borrowed from the Norwegian cabins they'd seen on their voyages. Swinging an adze to peel spirals of wood off the curved trunks and flatten the top side of the trees they'd gathered, they also brought in raw timber from storage on the ship. Notching the ends of the large bottom beams to nest one into the other at the corners of the house, they built a base four beams high in a recognizable log-cabin style. Since it was impossible to drive piles into the ground, aboveground pins were used in pairs to brace the building's corners and hold the stacked, notched beams in place.

They began to frame out the first timbers of their temporary home.[7] On one end of the stacked logs, they cut the top beam to support a doorframe. While they worked, they watched the current come and go. Some ice slipped out to sea, leaving an open space on the water. They looked at the sliver of open sea, then at their ship some distance away from it, still trapped and damaged. If they hadn't had to spend their days hauling lumber, they could've repaired the ice-knees and battered stern, preparing it for a homeward voyage. If the ship weren't locked in ice, the prevailing wind was just the kind that could carry them away from Ice Harbor.

But this was wishful thinking. There was no chance of escape at this point, and they had no faith that they'd survive the approaching winter. The sight of open water persisted into the next day, taunting them. They continued work on their house and burned scrap wood to try to keep warm.

On September 27, the wind whipped the air around their cabin. They paused, still wondering if some miracle might loose the ship and set them free. The frigid air affected them all. Holding nails between their teeth as they hammered planks, the men discovered that when they went to use one, the metal had frozen to their lips and tore the skin from their mouths, drawing blood. The ice drove in again. It was too cold to work on the house, but they had no choice if they hoped to survive.

They still had daylight left, but night was overtaking them. They'd learned to never walk alone and to travel in armed groups to protect themselves. On their way to the house, they ran into a mother bear and her cub, which they frightened off with shots.

The next day, a second group dragging their tools from the ship to the cabin saw three bears, which began to track them. They shouted and postured, but the bears didn't halt their approach. The humans felt much more vulnerable than the bears appeared to. But their mates already working at the cabin saw what was happening and set up a clamor, too, driving the animals away.

The castaways built a low ridge in the middle of the roof so that snow and water could run off, though for the time being, everything seemed to be made of ice. They knew they'd need a fireplace inside the house to survive, so a chimney was indispensable. Yet they could hardly imagine the strange uses to which it would be put during the winter.

On the night of September 30, the wind spun up a snowstorm so heavy that the crew couldn't fetch more wood. They thought to thaw the ground outside the house where they might pack earth around the base of the cabin to seal and insulate the building. But the ground turned out to be frozen harder than they'd realized. Wood was too scarce to burn in the quantities they would have to use to be able to dig in the earth.

Another storm the next day blew fierce enough to make it hard to move against the wind on shore. Snowdrifts rose everywhere, dropping visibility to a hundred feet or less. Yet by October 2, they'd completed framing the house. To the timbers they had nailed together, they added a May tree decoration made of snow. The next two days were spent in painful temperatures that grew too bitter to let them go outside on the first day, and so much snow fell that work became impossible on the second. Meanwhile, open water had crept its way closer to the ship. They secured the bower anchor on the ice to keep the vessel in place, in case the wind should try to pull it out to sea in their absence. On October 5, the sea lay wide open, except for where their ship was still frozen in ice, sitting two to three feet deep above the waterline. Below the surface, twenty feet or more of solid ice ran all the way to the ground.

They continued to work when they could, but eventually, the moment they'd been dreading came. Given that they were on an Arctic island without substantial vegetation, they'd found an inordinate amount of wood. The current from Siberia had pulled the timber up to the same shore for centuries, and they'd hiked miles to get it, dragging it wearily to the construction site over and over. But in the end, it wasn't enough wood to build a cabin. They began prying planks off the ship, taking them from the forecastle, the deck at the front of the vessel sitting a little higher than the main deck. Using these planks to cover part of the exterior of the house, they also laid them on the roof.

They hadn't removed the hull of the ship, or anything that would be impossible to restore come spring, if they made repairs. But dismantling any part of the ship was a final acknowledgment that in a choice between the ship and the cabin, their future for now lay snowbound on the icy shore.

On October 6, conditions were too harsh to permit work, so they stayed indoors. The next day, despite the cold, they took down the poop deck, which had been built up over the stern of the ship, and used those planks to finish enclosing the cabin. They caulked between the planks to seal them as best they could from the brutal winds and exposed terrain atop their modest hill. In the farthest reaches of the planet, they'd raised a house to live in.

Moving into the cabin was almost as much of a project as building it. On October 8, snow blasted their shelter on the exposed plateau, making it too hard to stand in the wind at all. It became impossible to walk more than a few feet outdoors. The following day brought no improvement, trapping them all in the ship again.

On October 10, the water rose two feet higher than normal, but the weather turned clear and allowed them to go to shore. They'd suffered for days in the choking smoke, sleeping between decks while the cook was forced to use his oven there, with the trade-off being that at least they could eat and be warm. Delighted to finally be at liberty, one sailor wandered off the ship.

He soon returned yelling, "A bear! A bear!" He'd been nearly taken by surprise and had been chased all the way back to the ship. The creature stayed on his heels until he got to the disemboweled bear they'd killed before and left frozen standing upright. The animal was mostly covered in snow by this point, but one paw protruded visibly in the air, and the live bear briefly stopped chasing its prey to examine the mysterious neighbor. Meanwhile, the fleeing sailor climbed into the ship, where the deck was filled with men crawling out of the hatches, smoke-blind and squinting. Trying to help, they were at first unable to even see the bear. Luckily, it wandered off before committing to climbing aboard the ship.

A break in the weather finally let them begin to move their provisions. That evening they shifted the bulk of their bread into a small boat and dragged it ashore. The next day, as they lowered the wine and barrels of food over the side of the ship, a bear—the same bear, perhaps, or maybe some new animal—rose out of the ice and began to move on them. It must have been napping, because they'd seen the lump on the ground as they came and went, but thought it was a piece of ice. They fired guns at the animal to spook it off, and the bear abandoned the hunt.

On October 12, eight men moved their belongings over to the house and stayed overnight in it for the first time. But they suffered terribly; they hadn't yet built bunks off the ground in which to sleep. More critically, though the sailors planned an opening for a chimney, they hadn't yet finished its construction. And no chimney meant no fire at which to warm themselves.

The following day brought savage winds. They began to move their barrels of beer, hoisting them over the side and onto a sled on the ice. But the wind spun up a storm that soon sent them fleeing back to the ship. The beer sat abandoned on the ice, while they huddled a few feet away, between decks. When they emerged the next day, they found the cask of spruce beer had frozen at the top. From there, the beer had kept expanding and burst the bottom of its barrel, freezing solid to the displaced panel as if glued to it. They loaded up the sled and dragged the barrel to their house, only to find the beer had separated. All the alcoholic spirits lay in yeasty unfrozen sludge, while the beery ice was little more than water. Melting the contents of the barrel, they tried to remix them, with disappointing results.

On October 15, the crew shoveled snow from the doorway of the house and laid stones to make a raised area outside the cabin in front of it. The next morning, when the men headed back for more provisions, they saw a polar bear on deck, which had apparently made its way onto the ship overnight. Whether the oncoming group unnerved it, or it thought better of its new den, upon catching sight of the sailors coming over the ice toward the ship, it slipped away.

Climbing aboard, the sailors began to take apart the cabin on the quarterdeck and carried the wood to shore to frame an enclosed porch for their shelter. An entrance hall would shield the interior door, keeping frigid air from sweeping through the cabin with every entrance or exit. Long attuned to the changeable nature of the wind, the men left openings for three exterior doors on the porch, one on each side of it. They'd be able to test the wind before going outside, and leave through whichever door would best shield them from the wind and snow.

It took two days to build the porch. On October 18, they finally unloaded the bread and wine from the small boat filled with temporary provisions they'd dragged to land weeks before, when they first feared the loss of their ship. Looking out from the shore, it dawned on them that the sea had finally frozen completely. They could no longer see any water at all. Water in its liquid form had vanished.

The same day, they spotted another bear, in what had become almost a daily occurrence. Some bears came after them; others were easier to scare off.

As with people, bears have personalities. Some reveal themselves to be persistent hunters, returning again and again to test the humans, while another bear might simply investigate and move on. Hunger likely affects the direction and depth of their investigations, making the bears more aggressive in dealing with humans.

Their roaming habits also reflect a certain variety. Using the sea ice as bridges, some bears cover geographical territories that stretch thousands of square miles, while others tend to stay closer to one location. Aside from the bears killed by the crew, which were permanently accounted for, the creatures seen by Barents and his men might have been a repeating cycle of the same animals or could've been new creatures just wandering through. In either case, polar bears clearly outnumbered humans at Ice Harbor.

Barents's men had already learned a lot about polar bears. In its natural habitat, each bear offered the same fusion of the mundane and the mythic as the Arctic itself. With its nonretractable claws and forty-two teeth, the animal possessed a lethal magnificence, yet the men had to find a way to live on a day-to-day basis with it. Polar bears weren't easily scared. They were used to the boom and crack and groan of the ice. They had no natural predators. There was nothing this far north that polar bears had to fear attack from—not reindeer, nor the seals they ate. Humans were new enough to spark curiosity but not yet familiar enough to spark fear.

In 1773, fourteen-year-old Horatio Nelson, later to become an admiral and hero of Britain, enlisted as coxswain on the HMS *Carcass* and set out on a polar expedition, trying the same route that Barents had taken almost two hundred years before. He got as far north as Spitsbergen, where he set off on the ice with a friend to attempt to capture a polar bear. He carried a musket with him and tried to shoot at the animal, but saw only a flash in the pan. The fuse burned but failed to ignite the gunpowder. He was unable to fire at the bear and had to be rescued by the ship, which fired a cannon to frighten off the creature. The bear, it's said, was on a floating piece that came away from the part that Nelson stood on, further keeping the ambitious, foolhardy teenager alive.

The lone child on Barents's ship didn't travel in any such honored role. But he'd get as much experience with polar bears as Nelson would find nearly two hundred years later—and perhaps more. On October 19, the boy and two men were working alone on the ship when a polar bear climbed the side of the listing vessel to force its way aboard. All three were unarmed. Faced with a raging bear making its way on deck, they panicked. They threw firewood at the animal, but it charged them in fury, sending the men scrambling into the hold. The boy, trapped on deck alone with the bear, climbed the rigging on the foremast. By this time, men on shore had heard the commotion and came running, firing muskets and driving the bear away.

As with all the bear attacks, once they ended, the crew had to go back to their duties. To make the decision to leave the ship even more bitter, the next morning, they caught glimpses again of open sea in the distance. Going below for more of their provisions, they found that the iron hoops on some barrels had frozen and shattered. Some of the barrels themselves had likewise exploded. The Dutchmen had dragged their alcohol all the way from Amsterdam but wouldn't have the consolation of drinking it.

Two days of clear, calm weather enabled a steady process of moving what still amounted to several months of rations. The sailors' small house slowly filled with casks and barrels. But on October 22, it snowed. The wind drove the snow harder as it came down, piling it all around the house. The eight men who were staying inside couldn't get out.

They'd hoped to move the rest of the crew the following day, and the sun rose to calm skies. But they feared the tempest would start up again, and the hike from the ship to the cabin loomed large for one of the sailors who was already very ill. On the twenty-fourth the rest of the company finally made their winter pilgrimage, pulling the sick man on a sled and dragging the hull of the scute—the larger of their two small boats—over the gritty, unforgiving landscape. Upon arrival, they carried their sick shipmate and left the boat outdoors, turning it over to protect it against the coming winter. They hoped it might be of use again come spring, if they survived that long.

Returning to the ship, they surveyed its plaintive hulk planted in a field of ice. Looking at the sweep of immobile sea, the last possible thing they could imagine at this point was that the waters would open up again and free the vessel. They picked up the spare anchor they'd laid out on the ice and coiled its cable back into the ship so that it wouldn't be lost under new layers of snow that would surely arrive. On the way back to the cabin, they carried the last of their food, leaving the ship empty of human life overnight for the first time.

They returned again the following days, climbing back into the ship, gathering supplies for the small boats, and clearing out the last of the equipment they wanted to bring with them. Once it was loaded onto the sleds, they got into formation and picked up the ropes with which to haul the sleds. Suddenly, Captain van Heemskerck noticed three bears heading from the far side of the ship toward them, lumbering over the frozen sea.

He shouted to sound the alarm for the crew and to frighten the bears. The sailors dropped the ropes they were hauling and looked around for any means to defend themselves. Two halberds lay on the sled with the equipment to be ferried over. Gerrit de Veer picked one up, and van Heemskerck seized the other. They began to protect the crew as best they could. Luckily, the halberd was a distance weapon, and didn't require them to fight within immediate reach of the bears. But with three bears, no guns, and two blades, they were overmatched.

While van Heemskerck and de Veer faced the bears on the ice, the rest of the crew made a break for the ship, where they could shut themselves in belowdeck. But as they ran, one sailor slipped and fell into a fissure in the ice and got stuck. Disheartened by the armed men, the bears turned toward the fleeing prey. De Veer was sure the fallen man was done for. But somehow the animals didn't notice or pay attention to the sailor on the ground, who lay mostly out of sight below the surface of the ice. They ran instead at those boarding the ship. The stranded sailor climbed out of the ice as van Heemskerck and de Veer caught up to him, and the three of them circled the ship, to climb it from the other side.

The bears, enraged, switched up their efforts to chase the smaller contingent. The crew picked up firewood and other stray objects, throwing everything within reach at their attackers. De Veer and van Heemskerck braced themselves with their lone halberds, realizing these weapons would likely be just as ineffective as firewood for killing the beasts.

The bears, however, didn't seem cowed by the assault. Instead, they grew curious and went to investigate the wood and other items that had been thrown at them, like dogs playing fetch. A sailor scrambled below to get pikes and bring back fire from the stove, but they couldn't start a fire quickly enough, making the guns aboard the ship useless to them.

When the bears finally turned on them for good, one man threw a halberd at the biggest animal and managed to split open its snout. Startled, the creature pulled back and realized its injury. After a moment, it began to flee, alarming the two smaller bears, which turned to run after it.

Once they were gone, the sailors took up the task they'd left off. In relief and exhaustion, the men harnessed themselves again to the sleds, which they hauled, bumping along without incident to the house. They'd moved entirely onto land, surrendering the ship to the sea. Not all of them would live long enough to return to it.

CHAPTER SIX

The Safe House

At Ice Harbor, the shelter Barents's men hewed out of materials they had at hand stood as alien as the Dutchmen themselves in the Arctic landscape. From a low spit of land almost level with the sea and covered in ice, a short rise led inland and up to a flat plateau of ground. On the higher ground, a rectangular wooden box, half log cabin and half cottage, sat facing the sea on three sides. The roof angled upward to a ridge at its center, and in the middle of the ridge, a small, square tower narrowed like a pyramid, ending with an empty barrel for a chimney.

Later explorers arriving in the high Arctic would embrace many styles of shelter. At Jackson Island on Franz Josef Land in 1895, Fridtjof Nansen and his partner Hjalmar Johansen dug a three-foot depression in the ground then erected a hut over the hollow out of moss and stones. Other kinds of housing appeared that were even less permanent. On Robert Peary's attempt to reach the North Pole in 1909, Inuit assistance meant igloos could be built to cache supplies and provide sleeping quarters, saving explorers the trouble of carrying camping gear.

By the time Nansen was trekking on foot closer and closer to the Pole, steamships were already ferrying tourists from Europe to Spitsbergen during summer months. Along the northwestern coast of the island, the construction of a tiny shack just a few feet wide called Lloyd's Hotel made it possible for adventurous aristocrats to have a drink in front of a miniature cottage and get their photos taken in a remote corner of the Arctic before returning to their ship.

Later, European-style builders would labor at other projects in the high Arctic. Less than two decades after Barents's mates cobbled together a cabin on Nova Zembla, whalers would visit Spitsbergen hundreds of miles to the west to hunt. Their worksites and sleeping quarters would evolve from tents to wooden barracks and low brick tryworks—furnaces for boiling whale oil—that they would return to year after year.[1] The sailors who were abandoned in 1630 while deer-hunting on the coast of Greenland as *The Salutation* fled approaching ice had borrowed tiles, timber, boards, and bricks from nearby tryworks to build a refuge while they waited for their ship to return the following spring.

Over time, some shelters would become more elegant. As coal mining took root in the high Arctic, accommodations were designed for company executives during their brief stays. Later, in 1926, Norwegian explorer Roald Amundsen would take over one such building on Spitsbergen, living in a two-story house at the harbor of Ny-Ålesund, where he waited to launch his airship *Norge* and drift aloft over the North Pole.

Several of these cabins would survive for a century or more, serving as partially or fully stocked outposts for hardy tourists or as lifesaving landmarks to re-orient lost explorers. Over time, several were transformed into research outposts. But before Barents's arrival at Ice Harbor, no shelter had ever been built so far north by Europeans. The only human architecture this close to the Pole had been constructed by indigenous peoples, who'd typically moved south over time in search of more resources and a kinder climate in which to thrive.

Forerunners of the Inuit who migrated to the coasts of northern Greenland centuries before Barents sailed, the Thule made skin tents in warmer months and dug out homes in winter. Before them, the Dorset culture in North America and Greenland built stone longhouses thirty or more feet long with hip-height walls but no roofs, likely pitching sleeping tents inside their rock enclosures.[2]

Artifacts of these civilizations would, in time, be unearthed and painstakingly interpreted. But deciphering the structure and use of William Barents's cabin would require far less work. Het Behouden Huys—the Safe House—as the Dutchmen came to call the structure, would stand largely intact for more than three hundred years.

By late October, when all the Dutchmen had settled into their new cabin, the large, long room was filled with provisions and equipment. Away from the door and off the floor, six raised bunks for sleeping, likely in rotating shifts, perched along one wall. A fireplace sat in the middle of the cabin, its smoke drifting up toward the chimney and barrel overhead. The sailor who had been too ill to transport for a time had room for a pallet near the fire, for warmth. The long guns leaned upright against a wall near the entrance, ready for use.

Not everything aboard the ship would fit in the cabin, and many items couldn't be easily carried to land, but the crew brought what seemed necessary and practical. They dragged their most valuable cargo: pewter plates and candlesticks, and a packet of expensive scarlet fabric. Inexplicably, they also preserved some less valuable items intended for trade, including thousands of pages of cheap prints in stacks, poor representations of the budding of one of the richest eras in art history. In time, they might return as many items as possible to the merchants who'd supplied them for the voyage, to show they'd been responsible stewards.

Yet the cargo they saved hardly represented their most important possessions. After their food, the most valuable items were tools and clothing, along with Barents's maps and charts. These were the things that might help them survive or to make their way home. They had any number of bladed weapons, from swords to halberds and axes. They had their guns. They had hammers, a dozen or more sizes of awls and chisels for piercing or gouging, a hand drill, a handsaw, and an adze for planing wood flat. They had wooden and metal beer taps for their barrels, a marking compass for measuring distances on maps and charting routes, and a shoe-maker's last—a smooth block of wood in the shape of a foot. They also carried with them a parchment-covered small book titled *The History or Description of the Great Empire of China*—though by then the promise of the Far East lay even farther away than it had when they'd set out from Amsterdam.

The first day that everyone sat together in the cabin, it grew too cold to work outdoors for long. But the men managed to kill an Arctic fox. They skinned the animal then roasted it, deciding it tasted like rabbit. The rest of the day was spent doing practical chores that let them stay indoors.

Except for the presence of human life in a place with so little to sustain it, perhaps the most extraordinary thing in the Safe House was the ship's clock. It reached knee height from the cabin floor, stretching as wide as the fingers of a spread hand and almost as deep. Framed in the shape and size of a lantern, it had an elegant cast bell on its crown. Made of forged iron, the clock had a round face with arrow clock hands, a winding key, and exposed sides, revealing the motion of gears marking orderly time in a place where the sun and sky seemed to change each minute. Along with the clock, the cabin's occupants had their hourglass, which they turned and tracked as another way to divide the twenty-four hours of their darkening days.

Everyone on the ship had just spent weeks together under a midnight sun. By logical extension, polar night would follow. But what would permanent night mean for the bears that stalked them? What about the animals that might keep the men alive when their rations dwindled—would Barents and his fellow sailors still be able to catch foxes? What would it be like to live in complete darkness once it came? With the ship's clock and hourglass, they would have a way to cross-check the passage of time, to throw a net of predictability over their lives. They could count the remaining weeks until the New Year, the sun's return, and the arrival of spring. Accurately marking the passage of time offered one way to keep some sense of control as the sailors braced for winter.

The day after the clock was wound and first struck the hour in their new home, the men set out to gather firewood for the long winter. The supply they'd found up the coastline as a prelude to building the house remained a reliable source of fuel. Driftwood would continue to wash up in the same place, but the site lay miles from the cabin.

The crew members began their hike, but along the way, a storm blew in, forcing them to turn for home. When the storm subsided that evening, three men ventured out to visit the bear they'd killed and left frozen upright. They wanted to extract its teeth. Despite its impressive height, however, the bear was nowhere to be seen. Wind and snow had buried it. While they searched for the posed carcass, the storm rose again, catching them by surprise. More snow came down in torrents until they lost all visibility and were walking blind. The snow in the air and on the ground and the sky in evening blurred the boundary between them. They staggered through the blizzard back the way they thought they'd come. Searching for the house, the sailors nearly missed it altogether, which would've meant death.

Other disasters loomed. The roof turned out not to be weather-proof; the planks of pine they'd pounded onto the frame didn't sit tight enough against one another to keep storms out. On October 29, they gathered pieces of slate from the beach, stretched the sail they'd brought to shore over the top of the house, and weighted it with the flat stones. The following day was clear, but on the last day of October and the beginning of the new month, the new roof was tested by a tempest that trapped them in the house.

The sun still made an appearance on November 2, half-rising and rolling along the ditch of the horizon before going down without fully revealing its face. Taking advantage of the visibility that remained, one of the sailors killed a fox with a hatchet, providing everyone a good dinner.

The bears that had terrorized them just a week before seemed to have disappeared. Daylight fled with them. The next day, only the tip of the sun was visible, the last remnant of a beacon that had lit their way for months. Barents took the height of the sun's thin top sliver to reckon his latitude. The next day it was gone.

As soon as all sixteen surviving sailors had moved into the cabin, Hans Vos, the ship's surgeon, imposed a health regimen for the crew. Water or snow dumped over hot rocks laid underneath a seat in the middle of an empty wine barrel served as a hygienic sauna. By the time each sailor took a turn in the steam bath on November 4, they'd likely not washed for some time.

Their filth was not unusual. During the Middle Ages, which French historian Jules Michelet sweepingly dismissed as "a thousand years without a bath," bathing didn't vanish as a practice, but it was often done on limited occasions for particular reasons. Bathhouses existed, but in the fourteenth century the Black Death had reduced their popularity. Rising Protestantism in Barents's century remained suspicious of the social license and sexual promiscuity attached to bathhouses' reputation.

But on land and aboard ships, baths were still believed to have their uses. In the eighteenth century, British captain James Cook— who sailed to New Zealand and Australia and died trying to kidnap the the ruling chief of the island of Hawaii—would prescribe cold baths for his men, along with other hygienic measures. Even in Barents's era, doctors sometimes asserted medicinal purposes for certain practices, treating the crew preemptively in the name of health.

But Dutch ship's surgeons suffered from poor reputations— reputations largely deserved. Many had trained as little more than barbers and ad-hoc dentists, getting minimal formal study. Ship's surgeons as young as thirteen appeared on the books, but more often were at least twenty years old when entering service as a junior surgeon on a ship. Yet even junior surgeons might find themselves in charge of all medical care if the first surgeon and surgeon's mates fell ill or died.[3]

With the risks of contagious diseases shipboard, the lack of expertise among ship's doctors could take a real toll. Decades after Barents's voyages, Dutch captain William Bontekoe wrote in his journal of surgeons who "wandered the high seas . . . like executioners," tormenting the helpless crews.[4]

Sometimes, however, surgeons did no harm, or even helped. During their first week in the house at Ice Harbor, the men enjoyed their steam baths. "It did us much good," wrote Gerrit de Veer, "and was a great meanes of our health." It's possible the men may simply have felt grateful for a few fleeting minutes of warmth.

The next morning, some of the sailors headed out to look at their ship. From a distance, they could see it still lay on its side, locked in its winter prison. As they trekked out onto the solid sea, the moon lit their way. With the sun's disappearance, the idea of day and night became less and less relevant. A dull predawn or twilight glow still lingered, diminishing by another few minutes each day. But the echoes of the sun dully lighting the sky from below the horizon became harder and harder to distinguish from the light of the moon, which sometimes stayed aloft and visible in the sky for a week at a time.

In a clear sky, the moon could light the entire landscape well enough to see for a mile or more. For practical purposes, night became the absence of the moon on a cloudy day, or its cyclical waning in the sky each month. But whenever its light was blocked, the treacherous Nova Zemblan terrain was smothered in darkness.

On November 6, the sailors managed to haul a sled of firewood to the cabin. The next day they realized that the clock had stopped, and they found themselves altogether unsure of the time. They had stayed in bed that day and hadn't yet gone out to relieve themselves in the snow. But after the clock failed, they couldn't tell by standing outside whether it was day or if day had turned to night.

With close observation of the sky, they eventually deduced it was around noon. But the clear routine that they'd lived on the ship—which both drove and was driven by the demands of the day—was gone.

On November 8, the crew dragged another sled piled with wood from miles up the coast, and trapped another Arctic fox. Open water in the sea caught their attention, but any return to the sea was months away or more; they'd committed themselves to the shore. They pried open a barrel and divided out their portion of hard bread for the next eight days, which came to just under five pounds each. Previously, a barrel had only to last five or six days. Now they began to think about how to make the provisions they'd set out with that summer last a full year or more.

The saddest discovery that day, however, was that their beer was already running short. Though much of it had spilled, and the barrels that had separated into yeast and ice when frozen didn't taste at all the same as before, the residents of the Safe House were nonetheless disappointed to find their ration cut.

In Barents's era, sailors drank nearly a half-gallon of beer each day in winter—and more in summer—for as long as their ship's stores lasted.[5] It was weaker than beer produced centuries later, but it did contain alcohol. Along with bread or ship's biscuit, beer was a source of B vitamins and calories. More than ship's biscuit, it was also a diversion, and a way to sustain life.

Sailors were suspicious of drinking water in general, and rightfully so. Though they didn't understand the dangers of bacteria, and the germ theory of disease had yet to appear, they understood that water sources in unfamiliar lands could make them sick. Mixing beer with water also killed the algae and bacteria growing in freshwater barrels that sat for months in the hold.

Alcohol in its various forms provided many sailors with their chief means of hydration.⁶ Trapped in the Arctic, Barents's men were lucky on this front—they could melt snow without fear of drinking it. But drinking large quantities of water was alien to sailors, except those in the most dire straits. Losing part of their alcohol rations meant that the dependable world was out of kilter.

On November 9, it grew dark. The next day, they hiked out to the ship to see how it lay on the ice. Opening the hatches and climbing down to the orlop deck, they peered the next level down into the hold. The space was filled with water that covered the stones used for ballast. But the water that had crept in through crevices and cracks had frozen fast, leaving the hull of the ship a diminishing barrier slowly losing the battle to keep the sea outside separate from the sea within. In any floating ship, even one with a hole in the side or irreparable damage, the answer was to pump the water out. But ice couldn't be pumped, and for the time being, they would have to let the sea possess their ship.

The sailors also turned their attention to matters more under their control. Foxes were good eating, but not always at hand, particularly when the men were spending most of their time indoors and had only moonlight or darkness to hunt by. To automate the process, they wove a round hoop of cable rope, knitting it into a net, and built a trap that would fall on any foxes creeping under it. They put out their trap on November 11 and were rewarded with a hot dinner.

Trapping in the Arctic was an art and a heritage, one owned almost entirely before and during Barents's era by native northern populations. The Thule people, forerunners of the Inuit, built fox traps in Greenland that had sliding stone doors and narrowed to just the width of the animal at the end, drawing them in with bait but leaving them no room to turn around and escape. Propped stones might drop a stone directly onto a trapped animal, crushing it. Or a suspended flat stone door would be tripped, sliding shut.

Traps for polar bears could work the same way. By the time the animal got deep inside, there was no room to turn and use its powerful forepaws to dismantle the opening. With any luck, before the trappers would come back, the creature had starved or frozen to death. But Barents's men weren't yet so proficient at trapping northern animals, and had to make do with the foxes, whose meat they preferred anyway.

The surplus fresh meat came as good news, but on November 12, the wine came in for rationing as well. They began to drink fresh water to stay hydrated, melting it daily out of snow. The men spent the rest of the week indoors without much to do. On November 18, the weather was bitter. Van Heemskerck, in charge of the merchandise for trade, pulled out a bolt of coarse woolen cloth from the cargo. Unfolding it, he began to cut it up and distribute it among the crew, who needed warmer clothes. The next day, he opened a chest of linen, from which they stitched shirts for themselves.

The day that followed revealed itself to be clear and calm. The sailors went to wash their clothes in a pot of water boiling on the fire in the cabin. As they lay out their shirts to dry, the edges of the cloth closest to the fire remained wet, while the far end froze solid, flat as a board, and couldn't be opened. The men had to throw their laundry back in the boiling water again just to separate the layers of fabric.

While the sailors were more at their leisure than they would've been aboard the ship, the cook had the challenge of melting enough water for everyone, chopping wood to keep the stove going, and cooking meals twice a day. On November 21, to share the burden, the rest of the men agreed to take over splitting logs, with everyone working in rotation except the leaders of the expedition, Barents and van Heemskerck.

The next day, the company neared the end of the store of cheeses that they'd carried from Amsterdam and off the ship. Seventeen blocks remained. After dividing one among themselves, they each received a whole one, to eat when they wished.

As winter came in, they saw more and more foxes. The sailors built new traps, in the hopes of taking advantage of the greater numbers. These were traps of a different kind: heavy planks weighted by stones surrounded by bit ends of spars. The traps tricked foxes into triggering the plank's fall, pinning the animals in place, while the spars kept them from digging their way out the side of the trap.

On November 24, a storm raged outside, and several men inside began to feel ill. Four sailors took turns in the barrel sauna before being given a purgative by the surgeon. The purgative emptied their bowels in a manner meant to cleanse them of toxic elements, but the trauma of the treatment was more likely to do violence to their systems than to provide any real benefit. Yet sailors sometimes took the extreme response of their bodies as evidence of the treatment's very effectiveness. As Gerrit de Veer wrote, the purgative "did us much good."

If they were seeking more good omens, they could find them in the four foxes they caught that day with their new plank traps. Improving on the design by adding a spring, they quickly caught two more of the creatures. But the weather never stayed clear for very long. On November 26, a blizzard struck. Snow blanketed the ground before climbing higher and higher up the sides of their cabin, over the base logs, past the steps, covering the doors, and eventually blocking all exits. Unable to leave and reduced to desperation, the men relieved themselves inside, on the enclosed front porch.

The next day, they built more spring traps and caught more

foxes. Low on salted beef, they were happy to supplement their aging rations with fresh meat. But November 28 brought another storm, pinning them in the house again. The next day, they wormed their way out of the cabin and used a shovel to dig out one door, but their traps lay completely covered in snow. They had to clear the planks to make it possible for foxes to even enter them.

Worse events were underway. They didn't know it yet, but they were already afflicted by scurvy. The signs of the disease that had appeared in some sailors during Barents's second voyage inevitably began to appear on this voyage, too: joint stiffness, loose teeth, and diseased gums. Symptoms of scurvy had been observed just four months into the prior expedition, and the sixteen sailors socked in for the winter at the Safe House on Nova Zembla had now been at sea for more than six months.

Ascorbic acid—vitamin C—is critical for building connective tissue, making its absence felt there first. A condition brought on by a deficiency in vitamin C, scurvy works quietly but inexorably, stripping the body of its ability to stay in one piece. As it worsens, the scar tissue over old wounds dissolves, bowels loosen, and bones grow fragile. Humans have no way to store ascorbic acid long-term, which means that scurvy wreaks havoc on even the strongest of bodies, starting with minor changes as early as two months without vitamin C.

Scurvy begins with lethargy, fatigue, and symptoms like those experienced on the 1595 expedition to Vaigach Island. Without access to vitamin C, the sailors would in time face swelling, jaundice, severe spontaneous bleeding, convulsions, then death, as their bodies slowly disintegrated. In the meantime, as their cartilage vanished, they'd begin to creak when they moved.[7] The symptoms would go beyond physical distress—depression and hallucinations would develop as the disease progressed.

Much to the crew's misfortune, Barents had set sail in the middle of a bleak age for scurvy. It had developed into a common ailment only once improvements in navigation allowed voyages to last several months. Between the fifteenth and the eighteenth centuries—from the first Spanish conquests in the Americas to the colonization of much of the globe—nearly two million sailors would die of the disease.

Yet its cure was already known by some. Mexicans treated Spanish sailors who arrived scurvy-ridden on their shores in the sixteenth century, giving them oranges, lemons, and limes. British explorer and pirate Richard Hawkins delivered hundreds of citrus fruits to his crew in Brazil, hypothesizing on their value fighting scurvy in a 1622 letter: "that which I have seene most fruitful is sower oranges and lemons . . . I wish that some learned man would write of it, for it is the plague of the sea, and the spoyle of mariners."[8] Hawkins claimed during his two decades at sea to have seen thousands of men afflicted with scurvy. Many individual captains from seafaring nations would, like Hawkins, bring along preventatives, yet others would fail to do so, leaving scurvy as a significant threat to sailors well into the nineteenth century.

Because little in the way of vegetation grew on the far northern end of Nova Zembla, even if Barents's men *had* understood the illness, they wouldn't have had any oranges and lemons, or even potatoes or broccoli, to save themselves. No plant they could find to eat in their corner of the high Arctic would cure scurvy. Most animals automatically synthesize their own vitamin C. Humans—along with bats, primates, and guinea pigs—are among the few that can't. But Arctic foxes can. Consumed fresh without being overcooked, the flesh of the foxes eaten by Barents's men carried small amounts of vitamin C. The fox meat was the key thing that was now keeping them alive—or at least slowing their death.

Without knowing it, their survival had become a race between the accelerating disease and how many foxes they could eat.

Meanwhile, trapping kept the men busy and gave them something to look forward to. They liked the meat and thought to make another use of the animals by skinning the creatures and turning their pelts into fox-fur hats, finally taking a step toward appropriate clothing for the climate.

Clear weather came in on the last day of November. Six armed men made their way to the ship at midday, to see if it had shifted position. Going belowdeck, they caught a curious fox and killed it but didn't see any bears. The next day's weather was much worse, unloading torrents of snow, blocking up the doors yet again, and covering the entire house. With no ventilation out the barrel chimney above the fire, the house filled with smoke, and the castaways were presented with a choice of freezing to death or suffocating from smoke. Solomon-style, they split the baby and went without a fire for part of the day, huddling in their bunks, but suffering through the smoke and enjoying the heat when it came time to cook meat for dinner. On December 2, the same dilemma presented itself, but they heated stones and carried them into their beds for warmth.

When they had no fire or lantern lit, the cabin was dark day and night, with the rank smell of scurvy-ridden men and fireplace smoke. Even in the quiet, the breathing of sixteen men could be heard, as could the groan and crack of ice as it shifted or shattered out on the sea. The sound was such that they wondered if the largest hills of ice they'd seen were being driven into each other.

With only intermittent fire, the remaining heat bled out of the cabin, which was far from airtight. First frost, then ice began to build up on the inside walls, eventually lining the interior—even the roof—in a frozen layer more than an inch thick. The intricate ship's clock, which had stopped more than once, finally froze altogether. They hung heavier weights on its gears, to try to make them turn. But the machinery didn't budge. For three days, the sailors stayed in their bunks, the sides of which were also lined with ice. They hardly moved at all, except to turn the hourglass, so as not to lose track of time altogether.

On December 4, the storms stopped. Realizing that digging out the doors would be a miserable ongoing project, they divided the task into rotations, with each man having a turn. Once again, van Heemskerck and Barents remained exempt from the chores. They were not only the leaders of the expedition but the likeliest members to be able to deliver the crew home next spring, if anyone could survive that long. The two men had become precious in a very practical sense.

The next day brought clear weather again, and the sailors grew ambitious, going outside to clear and reset their traps. But the hope and energy that came to them when the storm ended fled on December 6 in the face of more bad weather. Stung by the bitter wind that made itself felt throughout the cabin, they likewise dreaded the cold that crawled in with it. The sailors began to fear that they might freeze to death in their beds. They tried to make a fire, but despite the flames, the ice-filled cabin couldn't be heated. Their sherry ration, which they'd hoped might at least warm their insides, had frozen solid. They tried to melt it over the fire, and ended up with enough for each man to have a cup.

The foul weather persisted into the next day, and the men conferred on what else they could do to survive. Believing that they wouldn't be able to endure the bitter temperatures much longer, one of the sailors remembered the mineral coal they'd carried from the ship to the cabin. They were sure that it would raise the temperature in the room enough to keep them alive.

Coal burns hotter than regular wood, but it was known to smell awful—so awful it couldn't be used to cook meat without ruining the taste. So the men waited until evening to start a good-sized blaze with the coal. It burned so warm that they agreed to stop up the chimney and doors to trap the heat in the cabin as long as possible. For the first time, perhaps, since their arrival, they lay comfortably in their beds talking to one another as they began to nod off.

But those still awake began to feel dizzy, with the sailor who'd long been ill feeling the effects of the smoke first. Everyone suddenly realized they were all sick. Recognizing that it was a result of burning coal, the least affected among them jumped out of their beds to unstop the chimney and throw open the doors. The first man to open a door groaned and collapsed in the snow. Hearing the moan, Gerrit de Veer jumped from his bed and saw his shipmate on the ground. He grabbed vinegar and splashed it on the unconscious man's face. The fallen man woke and stood up. They opened the rest of the doors. De Veer realized that the cold air which had threatened to kill them only hours before now offered their only salvation. As the frigid temperatures invaded the house, the smoke cleared.

They'd nearly gone to sleep for the last time. If their hard-earned heat had now fled, at least they were alive. The captain doled out an extra ration of wine to commemorate the close call. The weather the next day remained vile, but cold as it was, they found themselves without any inclination to burn coal again.

On December 9, the wind and snow let up, and following their new routine, they cleared the doors and reset their traps once again. Two foxes stumbled in and were eaten with gusto—leading to more hats for the sailors. The eleventh found them facing such bitter temperatures that their leather shoes froze solid and could no longer be worn. Going barefoot wasn't a possibility, so the men carved wooden clogs to wear as slippers, attaching uppers made of sheepskin. When temperatures dropped even further the next day, frost crystallized on their shirts as they sat in the cabin, and icicles began to form on their clothes. They convened to debate building another coal fire indoors, but decided against it, fearing it would kill them even more quickly than the savage temperatures.

A warm day on Nova Zembla might approach thirty-two degrees, but in the winter months temperatures of minus thirty could be expected. At that temperature, ten minutes of under-protected exposure can trigger hypothermia. At ten degrees colder than that, five minutes might suffice.

No deep plunge in body temperature is required for serious harm to take place. As body temperature drops past three degrees below normal, shivering, weakness, and confusion set in, and the body diverts blood flow from the extremities to try to stay warm. As the body's core gets colder, amnesia and unconsciousness follow. By the time the human body reaches eighty degrees, death can occur.[9]

Once hypothermia takes hold, it's possible to revive a victim. But it must be done carefully. Hundreds of years after Barents sailed north, a group of Danish fishermen would be rescued from the North Sea after floating in water for an hour and a half. Brought to safety aboard another ship, they went to get a warm drink. Their body temperatures rose too quickly, and all sixteen died instantly.[10]

But if the temperature of the men in the cabin on Nova Zembla were to slip below the danger point, no one would stumble ashore and fish them off the coast. No one would take them aboard or resurrect them. Their cabin would sit undiscovered on Ice Harbor for centuries.

On December 13, they caught another fox and worked carefully to clear and reset all the traps. This had become a complicated labor. Staying outside more than a few minutes led to swelling and blisters on ears and faces.

Yet certain tasks had become reflexive. Though the sun was gone, William Barents went out and instead took the height of a star in the constellation of the Giant—Orion, the hunter and slayer of beasts—to reckon their latitude. The star representing the shoulder of Orion, either Bellatrix or Betelgeuse, told Barents what the vanished sun couldn't. Standing with a view of the snow leading down to the frozen sea on three sides, he recorded their position in the cabin at Ice Harbor.

Any thaw was still months away, and water sat higher in their icebound ship with every visit they made. Van Heemskerck and Barents already realized they might not be able to sail for home in the vessel that had brought them. With no rescue forthcoming, when it came time to set out, the men might have to take to the sea in their two small boats, with Barents at the helm of one and van Heemskerck captaining the other. Scurvy was already making its presence felt. They were sick and likely to get sicker before they left. As navigation historian Siebren van der Werf wrote, "It was important that not only they, but as many of the mates as possible . . . master navigation."[11] Barents likely repeated his measurements again and again with different celestial bodies to train the crew and give them practice. In the event that he or van Heemskerck died or became too sick to navigate, the crew would be able to carry on.

By December 16, they'd burned all the wood stored in the house, and they began to dig in the snow outside for the stack of logs that lay buried there. They went out in pairs in turn. Though no one could endure the cold for long, the crew had to shovel the high drifts all the way down to the ground to find wood. They hoped to save themselves from having to drag their sleds over the uneven landscape all the way to the driftwood shore and back. Even when the men heading out doubled up on clothing and wore their fox-fur hats, it was a painful chore, due to the "inexpressible, intolerable cold" and because the terrain around the house was covered in high drifts.

On the second day of fair weather, December 18, seven sailors made their way in darkness to the ship for the first time that month. Climbing aboard, they crossed the deck. The men hoped to assess the state of the vessel, and also to catch another fox. Just in case, they closed all the cannon portholes and openings on the ship to trap any hidden animal. Stepping onto the stairway to go belowdeck, they made their way into darkness.

No deeper night exists than in the hold of a cargo ship frozen into the Arctic ice during polar night. Looking into the hold, they lit a flame to see in the pitch black. The light revealed a fox, which they killed. Measuring the water in the hold, they found it had risen a finger length. But the new water, like the old, was frozen fast, and remained beyond the ability of the pumps to remove. The barrels full of fresh water, surely now frozen inside, were sitting surrounded by seawater and stuck fast to the inside of the hull.

Calm weather stayed with them for one more day, and they realized that the date was December 19. Though the longest days of darkness still lay ahead, they reasoned that with the winter solstice approaching, they'd soon pass the midpoint of polar night. With so little to console them on the frozen shore, they longed for sunlight, "the greatest comfort that God sendeth unto man here upon the earth, and that which rejoiceth euery living thing."

They caught another fox the next day, but bad weather that evening drove them back into the house, making the cabin into a tomb. They had a brief respite, during which they dug themselves out again and fixed the traps. But the following day a storm came, and they had to begin shoveling all over again. Months more of the same drudgery lay ahead. But on December 23, their spirits rose again. Even the simplest sailors understood the sun was in the south. Barents knew that it was shining its warmth directly over the Tropic of Capricorn, thousands of miles away. But the men in the cabin also knew that it was at the farthest point on its arc away from them. Each passing day would bring it closer.

After hearing the ice crash and crack out at sea, they woke to fair weather on Christmas Eve. Opening the door, they noticed that the sky was clear and the moon shone so brightly, they could see all the way to open water. Visible trails between sections of ice beckoned, but they might as well have been hallucinations. As evening came on, a storm spiraled in, burying them alive once more.

On Christmas Day, there was little to do but sit inside and listen to the wind and snow. Drifts had piled ramps of packed powder up the sides of the house. Despite the noise of the tempest, the sailors heard the scrabble of footsteps overhead. Any initial panicky fear of bears gave way to reality. A bear would've been a heavy presence on their fragile jury-rigged plank-and-sail roof. Their visitors were surely only foxes. Still, some of the more superstitious men announced it as a bad omen.

The moon stood in the twenty-sixth day of its cycle in the sign of Scorpio. Like the footsteps on the roof of the Safe House, star signs could be interpreted as superstition or science—and sometimes both at once. During Barents's first decades at sea, Belgian astrologer Cornelius Gemma would try to use the stars and divination to understand humanity's place in a godly cosmos. Yet he would also be among the earliest to observe a supernova in 1572 and to identify the location of a comet that passed over the Earth in 1577. (He was also the first to draw the humble tapeworm for posterity.[12])

Like many contemporaries of their era when confronted with the uncanny, the Dutchmen trapped on Nova Zembla for the winter tended to focus on the practical even when they remained respectful of a supernatural influence on events. While some of the sailors in the cabin feared the footsteps on the roof, others took issue, wondering why they should mean anything bad at all. They resolved that it was a bad sign only in that the weather was too rough for the crew to be able to trap and cook the trespassing foxes, because it would've been much nicer to have the creatures in their stomachs than on the roof.

The storm continued as they tried every strategy they could think of—building a wood fire, piling on layers of clothing, and heating stones and cannonballs for their beds. On December 27, they sat before the fire as close as they could, scorching their shins. Yet their backs remained covered in frost, reminding them of Dutch peasants who would arrive at the town gates on winter mornings riding sleds after traveling all night. For three days, between the storms and the cold, none of the cabin dwellers dared to go outside.

On the fourth day, one intrepid sailor dug a small tunnel in the snow in front of a doorway and went out to take stock of their situation. He came back in fear of losing his ears from cold, telling them that the snow had surrounded the house and had piled up as high as the roof. They were completely buried.

On December 29, seeing that the storms had stopped again, the next man on shoveling duty dug a tunnel out and cut a set of stairs into the packed snow so they could climb up to get to air. They hadn't taken any foxes in days but set about clearing the traps, and discovered one dead animal for their trouble, frozen stiff and not at all decomposed: a gift.

The following day came and wiped out all their work, with snow piling even higher overhead. The next morning brought more. They felt like prisoners, and when they sat before the stove to warm their feet, their stockings began to burn before they felt the heat—only the smell of scorched fabric alerted them to the danger. Barents's men spent the last days of 1596 crouched by the fire in doubt and misery, patching holes in their stockings and imagining months ahead in which they'd be buried in their cabin and resurrected over and over in eternal darkness, a little closer to their final deaths each time.

CHAPTER SEVEN

The King of Nova Zembla

A fter that with great cold, danger, and hardship, we had brought the yeare vnto an end," wrote Gerrit de Veer in his journal. "We entered into ye yeare of our Lord God 1597, ye beginning whereof was in ye same maner as ye end of anno 1596 had been."

The landscape held no relief in store, with snow, tempests, and always the cold haunting them. They spent New Year's Day trapped in the cabin. Offered their ration of wine, a small serving every other day, some of the men abstained for the time being, imagining how much hardship lay ahead and thinking it might be better to bank their luxuries against some future need.

No light appeared except moonlight and what they created with fire. The ice on the inside of the walls muffled outside sounds, but the smell of humans trapped in tight quarters remained sharp enough. The scurvy spreading through the crew would lend its hallmark stench to unwashed clothes and bodies, making the cabin even more oppressive. Worst of all would be the urine and feces piling up at one end of the shelter.

January 2 unfolded much as New Year's had: trapped inside again, they were reduced to using the last of the firewood they'd gathered from around the house. There was no question of getting more in that moment; it wasn't possible to be outside for any length of time and survive. Instead they began to pry off non-essential parts of the doorframe. Next they chopped up a wooden block on which they usually prepared their fish when it came out of the barrels.

Not many superfluous things remained to burn in the cabin, however. And the following day was no better than those that had gone before. No matter how careful they were with their wood, the supply dwindled again to nothing.

A sailor's destiny was tied to the weather and the wind. Along with a ship's location, these were often the first details that went into the log after each date, even when sailors were ashore. But by January 4, conditions grew so bitter that they didn't dare to open the doors for even the seconds it would take to determine the direction of the wind for the log. Instead, they hoisted a bit of scrap tied to one of their half-pikes, pushing it up through the chimney and watching to see which way it blew. Even that was tricky—they had to notice which direction it rose as soon as the pike went aloft. Within seconds, the separate parts of the impromptu windsock froze into one solid staff.

It was fitting that they continued to record the wind direction day after day, because they were still very much at sea. Trapped in their cabin like a ship battened down in a storm, their barrel chimney in place of a crow's nest, not only had night crowded out day for them, but space and time had changed places, too. They no longer moved through the water, cresting wave after wave and trying to stay afloat; their shelter had become stationary. Yet a whole sea of weather was borne aloft in the sky and set in motion around them. Just as they'd gone in search of safe harbors along the way, the men now waited for days of calm weather in which to disembark from their cabin. As in unknown waters, they had little idea what might come next, or how they would survive long enough to find themselves warm and dry, or when they might be able to leave their temporary quarters to walk on land and hunt for provisions.

At sea, William Barents had been the expedition leader, setting the course and keeping them to it, watching the sky and sea and interpreting them. The men trusted in his knowledge and accepted his vision for a route to China.

They'd come as far as they could. Their dream of China and Cathay was gone. The knees of the ship were damaged; the timbers were leaking, and most of winter still lay ahead. Rather than crossing innumerable miles and unknown seas to find their way to the Far East, they were now forced to move entirely through the medium of time. If they lived to reach daylight's return, they would've attained the only goal left for them to pursue. They would make a run for home.

If they survived long enough to leave, Barents's role was to offer that earlier vision inverted. Just as they'd once believed he might lead the way to China, they now understood his role would be to deliver them back to the Netherlands. Van Heemskerck still had charge of the men—the sailors reported to him. But Barents's role in the daily affairs of the house had become largely ceremonial. He would read and interpret the sky until such time as the currents and weather thawed the sea and set it in motion again.

On January 5, calmer weather arrived. They got to work while they could. They went out for wood and scooped out their waste from the house. They cleaned and straightened up the cabin as best they could. Once again, the house lay covered in snow up to the roof. Anticipating that they'd soon become captives again, they removed one of the three doors to the porch and dug a vault in the snow outside the porch to serve as a latrine.

Near the end of their workday, they remembered that it was Twelfth Night, the eve of the feast of the Epiphany, which celebrates the long journey and arrival of the three kings, or Wise Men, who followed a star to the Christ child twelve days after his birth. Epiphany itself was a holy day in the Catholic Church, but it hadn't been included in the shortened lists of religious holidays in the Netherlands. Yet the Twelfth Night feast was both religious and secular—and remained the most important Dutch family celebration of the year.

At home in Amsterdam, people would be out in processions singing and carrying lit paper and candle stars, stopping at houses and asking for gifts. Others would gather in taverns or homes to draw paper lots assigning them a role. Depending on the lot they drew, partygoers would fill any one of a number of identities for the night: pourer of drinks, servant, gatekeeper, jester, cook, confessor, queen, or king. Servants and peasants and burghers might sit down together. Children would jump over lit candlesticks, sometimes a triple candelabra, one flame for each of the kings. The poor would be fed, and charitable donations made.

Anyone who drew the right lot or ate a piece of cake that had the lone ceremonial bean baked into it might be king. He or she might be a servant or a master in regular life, an adult or a child, but as king, the lucky partygoer would spark the feast into motion by taking the first sip while the room cheered, "The King drinks!" Wearing a paper crown, often bearing medallions, the king would be in charge of the evening. If the celebration happened at a tavern, he might have to foot the bill for everyone.[1] All customary social interactions would be upended for one night, as the world turned upside down.

When the Nova Zemblan castaways realized it was Twelfth Night, they asked their captain for permission to celebrate with the kind of feast they would've had back home. Van Heemskerck agreed, and the sailors opened two pounds of ground meal that had been intended as glue to make paper cartridges for their firearms. The ground meal instead became pancakes cooked in oil. As a bigger treat, each man got one of the captain's wheat-flour biscuits. With several men having saved their ration of every-other-day wine for days, enough remained for everyone.

That night, they made a banquet to celebrate the holiday and drank to the three kings, who were also explorers led by stars into unknown mysteries. And in the European tradition, they distributed tickets and drew lots to see who would be king. The ship's gunner won the honor, becoming more privileged than his shipmates and more honored in the cabin for the evening than William Barents or Jacob van Heemskerck. For a single night he became king of the feast, and king of Nova Zembla as well, monarch of a harrowing, inhospitable land in which they feared they would die.

They weren't the only expedition to throw a party in the face of the unknown. Sent out into Northern Canadian waters in search of the lost Franklin expedition, the whaling barque *George Henry* got trapped in ice in the winter of 1860 but celebrated Christmas optimistically, with a feast and a gift exchange in which Captain Charles Francis Hall gave his Inuit guide Taqulittuq a Bible provided by the

Young Men's Christian Union of Cincinnati. Stranded in their small cabin on Franz Josef Land at New Year's in 1896, Fridtjof Nansen and Hjalmar Johansen would turn their clothes inside out to welcome new life and a new year, finally agreeing to stop using formal titles of "Mister" and "Professor" with each other more than two years into their travels together. For centuries, Europeans tried to re-create the pageants of their habitable world in a North that regularly thwarted their dreams of easy passage.

But the feasts, improvised celebrations, and gifts couldn't always overcome the longing for home that they evoked. Sailing with Nansen aboard the *Fram* before their inside-out New Year's Day, Johansen wrote of the crew's celebration over the course of their three-year voyage. For their first Christmas in 1893, Nansen showed up at dinner with small gifts from home for everyone: knives, cigarettes, pipes, and a dartboard. Their beer had lasted until Christmas, and they had cakes, almonds, and raisins to go with their meal. But by the following year, their high spirits and their beer had both vanished. Despite the invention of a baking-powder-and-cloudberry-jam "polar champagne," Johansen noted they were all too aware of the time and distance away from home to enjoy any celebration.[2]

> It seemed however, as if the true festive spirit was wanting, for this Christmas was not a very lively one . . . we were well and warmly housed there in the ice desert, but we were prisoners. We lay, far away from the world, fast in a frozen sea, where all life was extinct.

By the time the heroic age of explorers arrived—in the age of transatlantic telegraph and airplanes—the Nansens and Amundsens of the era knew that their exploits would be recorded, that whether they succeeded or failed in their polar journeys, whether they lived or died, they'd become legends. But Barents and his men lived without that promise. No one knew where they were. They had no idea what would happen to them. As they sat in the cabin on Twelfth Night with their frozen clothes, their three cups of wine, and their new king, they remained as close to mortality as any humans on the planet.

Barents and his men gathered atop the earth and stone floor, not far from the residue of their filth, drinking the dregs of their wine. They couldn't yet know who among them might or might not survive the winter—or even the night. But they sat toasting the three kings, who traveled far from home and witnessed wonders never seen before by man.

CHAPTER EIGHT

The Midnight Sun and the False Dawn

The morning after their celebration, Barents and his men returned to the monotony of survival. Making use of a break in the harsh weather, they cleared their traps and dug a second vault—this one for firewood. But then the storms resumed, which frightened them, because they weren't sure how much longer they could endure the worst of the weather they'd already experienced. Nevertheless, the sailors took heart the next day, when they began to see a glimmer at the horizon. The sun remained out of view, but its rays began to send up a whisper of dawn at the edge of the world each morning.

Yet grimmer portents persisted. One shipmate had lain in his pallet by the fireplace for weeks. Each of the sailors was getting sicker, their bodies wasting in starts and surges. Scurvy afflicted all of them. Its outcome, based in chemistry, was inevitable: in time, the lack of vitamin C would kill them. And those who had other illnesses or infirmities due to age might succumb more quickly; Barents, for example, was decades older than many of his shipmates. His health would be a fraught topic for everyone. Van Heemskerck and de Veer had sailed with Barents on his second Arctic voyage—the trip to Vaigach—but not the first. Which meant that not only was Barents their navigator, he was the only man among them who'd sailed the western coast of Nova Zembla all the way from its northern end to its southern tip—the route they'd surely have to take to go home again.

On January 10, a company of sailors hiked to the ship in clear weather. They went out heavily armed. They hadn't encountered any bears during polar night, but neither had they spent much time outside. Climbing up the side of the ship, they found round, fat footprints from several bears, adult and young, that had been haunting the ship in their absence. Opening the hatches, the sailors went belowdeck and lit a fire so that they could take a candle down into the hold. With the threat of bears and darkness overhead, they stopped to measure the new ice that had formed and realized it had risen a foot since their visit three weeks before.

The temperature the following day was less bitter, so they ventured a mile to a rise where they found more stones suitable for heating their beds. The next night showed itself clear and starry, and the men went outside to spot the constellation of Taurus in the sky and record the height of the star called Aldebaran. From there, they again reckoned their latitude. They had less than three months until spring.

The air felt even warmer on January 13, and the crew ventured outside long enough to try to play with the round ball from the top of the flagpole that they'd brought over from the ship. The next day they caught two foxes. A contingent returned to the vessel again on the fifteenth, finding that a sailor's jacket they had used to plug a hole on a prior visit had been dragged out by a bear. After months at sea, the coat likely smelled heavily of its owner, yet the creature that discovered it found no meat despite the whiff of prey. The coat had been torn to pieces.

A belief that exercise could stem or cure scurvy was common, so the men went outside the next nice day to run and play ball again. Though it was still winter, even the light from below the horizon seemed to warm them. Chunks of ice slid from the roof and sides of the cabin, and ice on the inside of the wall anchoring their beds likewise began to thaw. Yet all the warmth dissipated in darkness, and ice formed again.

Fire remained their means of day-to-day survival, but they seemed to go through wood faster than they could scavenge it. Driven once more to desperation by the cold, they agreed that they should try to burn their mineral coal again, as long as they didn't stop up the doors and chimney, trapping the fumes that nearly killed them before.

But they were also crafting another plan—one that made them hesitant to burn the coal. If they did end up having to try to sail for home over arctic seas in their small boats, they couldn't cross the open sea but would have to hug the coastline all the way heading north over Nova Zembla and descending hundreds of miles to its southern tip before making a jump toward the mainland.

Any plan to take the small boats would mean approximately eight men in each craft. Space for anything else would be tight. Their coal burned hotter and longer than wood and took up less room. Departure surely seemed like a distant dream in that moment, yet they had to imagine it, and hope. Banking on finding other ways to survive the winter, they decided to save their coal.

They'd have to find a way to endure other hardships as well. They had counted their barrels of bread and been opening them at set intervals. But some, they discovered, weren't full, meaning rations would have to be cut. An extra barrel of bread had been left on the ship and counted in the tally for the remaining days of provisions. But sailors hiking to the ship had been surreptitiously taking biscuits out of it over time, making the count even shorter.

Under a cloudy sky on January 20, they stayed in the house and broke up empty barrels to burn for heat. The foxes, which had been plentiful during their first weeks in the cabin, had become scarce. Perhaps it was just that the smart ones had learned to elude their traps, and the less intelligent ones had been eaten. But the sailors also recalled that the foxes had appeared in earnest once the bears began to vanish. The departure of their food source made them fearful not only about their dwindling supply of fresh meat, but also the prospect of returning bears.

Throwing a ball on January 22, some men began to believe that the sun was so close to the horizon that daylight would soon arrive. But William Barents said that it was still too early to expect the heavens to realign, that weeks remained before they'd see the sun. Gerrit de Veer accompanied three other men to the ship, where they thanked God for their survival thus far, talked of going home one day, and pried the lid off the bonus barrel to steal more of the ship's biscuits.

De Veer and van Heemskerck headed out with a third man on

January 24 to see the view of the sea facing south toward Nova Zembla. While they were making observations, they glimpsed the edge of the sun's disc slipping just above the horizon. They could hardly believe their good fortune. In fact, it seemed impossible. Barents had said that weeks remained before the sun's return. They rushed back to the cabin with the story of what they'd seen. Barents, too, was puzzled. Though the ship's clock had stopped, they'd kept time with their twelve-hour glass and checked it against celestial observations. How could they have managed to lose two full weeks?

It seemed so impossible that the sun would appear in their far northern latitude on that early a date that Barents declared the men mistaken about what three of them had seen with their own eyes. De Veer and van Heemskerck didn't back down, relying on their own account, though they realized that it was somehow "contrary to the nature and roundnesse both of heaven and earth." The crew began to wager on whether they'd see the sun in its glory the next morning.

But clouds or haze blocked the horizon the next two days, making it impossible to tell whether the light on the horizon was only a predawn glow, or if the sun had really returned. Meanwhile, they faced more immediate concerns. They spotted a bear coming from the southwest toward the cabin—the first bear they'd seen in months. Making a racket by shouting, they drove the animal away without a fight.

On the second of the two hazy days, the man who'd lain ill in bed by the fire for months grew even frailer. As night came on, they tried to comfort him, but whether scurvy opened the door to another affliction or was by itself enough to end his life, they had no cure. Just after midnight, in the early hours of January 27, 1597, he became the first shipmate they'd lost since the carpenter died. Fifteen sailors remained.

While the sky stayed clear, they sawed at the ground to chip out a grave for their companion. It was nearly impossible to dig, and in spite of the calm weather, the air froze their lungs and skin. They worked in shifts, to allow those outside to come in and warm themselves by the fire while others went out to take a turn. When they'd excavated seven feet of earth from the unforgiving land, they improvised a funeral service. Reading prayers and singing psalms over their mate's body, they lowered him into the hole they'd made. Returning to an emptier cabin, they ate breakfast. They honored their time with the dead man and commemorated his life, but his name is lost to history.

Their carpenter had died late in September just as the sun began to vanish below the horizon; the next sailor had left them just as they waited for it to reappear. In between, they had learned ways to survive, but they also realized that they would remain in danger as long as they stayed on Nova Zembla. At their funeral meal, they contemplated how to survive if storms blocked their doors and covered the house over and over, as had happened thus far. They struck on the idea of climbing up and out the chimney instead of using the porch doors to exit the cabin. On the spot, van Heemskerck left his dinner to try the new route, pulling himself through the roof of the cabin and out the open barrel.

The men inside the cabin suddenly heard him cry out. As they ran outside, they looked up and saw the sun. The sailors stood confounded by the view. Before them lay not just the thin line of a narrow blade of the sun creeping into view, but a whole celestial disc sitting above the horizon. The sight made no sense, because it broke the dependable laws of the heavens, which had been used to sail the globe for more than a century. Yet there was no mistaking the presence of what they'd all waited months to behold.

Barents had no answer for it. But they agreed that "seeing God is wonderfull in all his workes, we will referre that to his almightie power." The men would continue to count off the times they had turned the hourglass and recorded the passing days, and think such a thing was impossible. But they'd all seen it. They couldn't reckon it as either sense or nonsense, and in the end, simply described what they saw and stood by it.

In February 1894, almost exactly three hundred years after de Veer's sighting, Norwegian explorer Fridtjof Nansen would witness the same phenomenon. He was also shocked, but likewise confident in what he'd seen, describing it in *Farthest North*, his memoir of the voyage.

> *We had not expected to see [the sun] for some days yet, so that my feeling was rather one of pain, of disappointment, that we must have drifted farther south than we thought. So it was with pleasure that I soon discovered that it could not be the sun itself. The mirage was at first like a flattened-out glowing red streak of fire on the horizon; later there were two streaks, the one above the other, with a dark space between; and from the main-top I could see four, or even five, such horizontal lines directly over one another, and all of equal length; as if one could only imagine a square dull-red sun with horizontal dark streaks across it. An astronomical observation we took in the afternoon showed that the sun must in reality have been 2 ° 22' below the horizon at noon.*

In the polar regions, where the sun disappears entirely for part of the year, it's possible for an inversion layer, where warmer air is trapped above cooler air, to generate a mirage that's both real and unreal. If the inversion layer stretches uninterrupted for hundreds of miles, and the rate of temperature changes inside the inversion hit just the right window, sunlight can bend along a tunnel in the atmosphere, refracting sunlight. The real sun sitting below the horizon can be refracted over distance to appear in a shimmering, distorted shape above it.

Nearly four centuries would pass before the mystery of the prematurely appearing sun would be unlocked.[1] Using the Russian name for Nova Zembla, the mirage would come to be called the Novaya Zemlya effect, commemorating the location of the early sun that Barents and company saw in January 1597.

In documenting the atmospheric event, Barents and his men had passed even further into the realm of pure science. Their mission as merchants had been suspended, but nearly every observation de Veer made, down to the smallest references to plants and animals and weather, provided a wealth of information in deciphering the Arctic.

Beyond their mundane observations in 1597, as well as the wonders in the heavens that they couldn't explain, the Dutchmen were simply grateful to see the sun again in any form. On January 28, they went outside again to throw the flagpole ball and run, to to try to work off the lethargy imposed on them by their confinement and their scurvy.

Hemmed in by snow the next two days, they dug a short distance from one door. But mostly they stayed inside until fair weather on the last day of January brought them back out to clear the house and set their traps again. Spying a bear heading in their direction, they slipped discreetly inside and let it get close before they shot it nearly point blank. The blast spooked the bear but didn't kill it, sending it skittering away over the snow.

The first days of February were spent battened down in the cabin during storms, the sailors berating themselves for thinking that the early sight of the sun would mean warmer temperatures. It wasn't only a philosophical reflection: counting on milder weather meant that they had cut short their usual foraging for wood. Now they once again found themselves out in the stinging cold, rooting in the snow near the cabin in search of stray fuel.

When ugly weather socked them in again the next day, the sailors finally followed van Heemskerck's earlier example, surrendering the porch doors to the new-packed snow that kept them from opening. Instead, they climbed out the barrel on the roof. Those who were too sick to hoist themselves out through the ceiling were forced to relieve themselves indoors for four days in a row. The sun's early visit came to feel like trickery. Trapped inside the cabin with the ceiling their only sky, they had even less light than before.

On February 9—when they'd expected direct daylight to make a return—they felt the warmth of the sun, and their claustrophobia began to ease. During a string of days with good weather, they heard foxes on the roof once more, but also saw another bear. In their weakened state, they had no intention of fighting the animal, but only hoped to find a way to shoot any bears at close enough range to guarantee a quick kill.

Things had calmed enough by February 12 that they went out to clear the fox traps, which drew the attention of yet another bear, this one more curious than other recent visitors. As it came toward them, they slipped once more into the house to see if they could tempt it toward the entrance. When it reached the doorway, they shot it through the heart, the bullet driving through the breast of the animal and coming out its backside as flat as a coin. The bear recoiled from the blow and turned to run from the cabin. But before it got far, it fell to the ground. As the sailors approached its body stretched out on the snow, the creature lifted its head "as if he wanted to see who had done this to him." Unwilling to battle the bear even in its weakened state, they shot it two more times until they were sure it was dead.

The crew slit open its belly and gutted it, then dragged the body back to the cabin, where they skinned it and pulled a hundred pounds or more of fat from its belly. They melted the fat down for oil, which gave them enough of a supply to burn a lamp all night—something they hadn't been able to do for lack of oil. The slaughter of the bear provided enough for every man to light a lantern in his bed for sewing or writing or just for pleasure. The gift might have been even more appreciated in polar night, but the men took advantage of it all the same.

On February 14, they went to visit the ship and found little had changed, but the water in the cargo hold continued to creep higher and higher as the sea took greater possession of the ship. The next day found them stuck in the cabin once more, but they heard the tapping of foxes' footsteps outside, gathering to scavenge the corpse of the dead bear. Only then did it occur to them that the carcass might draw another predator, one that might catch them unawares. They vowed that as soon as the weather cleared, they'd bury the bear deep in the snow. But the following day was just as bleak, dumping snow once again. Realizing it was Shrove Tuesday—Fat Tuesday, a day of celebration and indulgence before the sacrifices of Lent—the men had a glass of wine each in their "great griefe and trouble" and pretended spring would come again.

The sky was clear enough to go out on February 17. They dragged the carcass of the polar bear out from where the foxes had been eating it and heaved it into the vault for their wood, which they'd since burned up. Once it lay in its snowy grave, they filled the opening and stopped it up, in the hopes of keeping the smell from traveling—a strategy unlikely to succeed, given polar bears' ability to smell prey thousands of feet away. They cleared their traps once more and again visited the ship, this time observing so many comings and goings in the clawed tracks that they realized polar bears had made it a way station.

The next night, they lay in the lamplight and again heard noises on the roof. This time, the crack and crunch of ice they were hearing sounded less like foxes and more like a bear. They listened with dread. But when they went up on the roof to scout the area the next morning, all they saw around the chimney were the footprints of foxes. They'd grown more fearful over the long winter and managed to spook themselves.

For some time, they hadn't been able to take the height of the sun, needing a clearer view in the sky, a clean line at the horizon, and enough sunlight to shine through the hole of their mariner's astrolabe. They improvised by making a quadrant, using two sights set in a line, and a lead weight suspended by a string. Combining the height of the sun with its declination from the equator, they again recorded their latitude.

February 20 arrived with brutal force, reminding them that winter wasn't gone yet. Wind and snow worked their fury around the cabin the following day, much to their despair. They had no wood left to burn. The sailors hunted the floor for loose scraps and broke off more wood from the interior of the cabin.

When the sun rose in clear skies on the twenty-second, eleven men set out with guns, blades, and a sled to hunt for more driftwood. But the inlet where they normally went lay so deep under the snow that there was no way to find wood, and they had to walk, weak and cold, farther up the coastline to find anything at all. By the time they managed to scavenge a few logs, they felt hollowed out by the trip but had little to show for it. They wondered to one another whether they'd have enough strength to even make the trip again. As they drew close to the cabin, they looked out at the sea and saw open water for the first time since Christmas Day. The thought of leaving would have to sustain them for now. They went back inside and began again the monotonous, unbearable task of survival.

The next day they caught two foxes, unwittingly delaying the effects of scurvy once again. Without vitamin C, scurvy can kill a human being in as little as five months.[2] More than nine months had now passed since the cabin's residents had set out from Amsterdam.

On February 24, they emerged under a dark sky and reset the traps in hopes of repeating the previous night's dinner, but caught nothing. After another day bottled up inside during a storm, they came out again to gloom and went through the motions of exercising. Again, their wood had vanished. They hardly had the heart to hunt for fuel, but on March 1 they prepared themselves and set out with the sled once more. Some of the sailors were too sick to join the group. One who stayed behind had lost his big toe to frostbite.

They managed to haul their sleds up the shoreline that day, but decided that going forward, they'd ration their wood, to limit the number of agonizing trips they had to make. Heating stones for the sick, they stayed in their beds and tried to get through the day with just one good fire during the night. Sometimes this was hard, but on other days, conditions became tolerable. Better weather on March 3 even led some of the sickest men to feel better and sit up in their beds to pass the time. But they soon found that their exertions took a toll and only made them feel worse.

Meanwhile, polar bears had returned in full force. A bear wandered up to the cabin the next day and got shot for its trouble, but the animal managed to escape alive. The same day, five men went out to visit the ship and found the now-expected signs of bears taking up residence in the vessel, but also something new. The bears had been busy, ripping off the hatch over the galley from under the snow, dragging it off the ship and over the ice.

They had to dig their way out again on March 5, but their reward was the sight of more open water flowing in the distance. As if to mock any hope of what it meant, a storm buried them yet again the following day. But climbing out the chimney, they once more spotted the open water, which seemed to be almost everywhere. The ship sat in the same pocket of ice where it had lain for months, still frozen in. With the storms that kept coming in and the winds that drove them, Barents and his men began to fear that they'd wake up one morning to discover that their ship had drifted away.

The haze and fog lifted a few days later and let them see even farther out to sea. Though there seemed to be a clear path through the water the way they'd come, ice and snow still packed the route to the east and southeast—in the direction of China. They dug out the doorways and cleaned their waste from the porch again as they discussed the possibilities. If only the ship could slip loose, the sea looked as if it might carry them away that day. The small boats they had on shore seemed so much frailer than their big yacht. But even if they were willing to consider the idea, it was still too cold to leave in the open boats.

That evening, they ran out of wood yet again. Nine sailors dragged a sled to the ship, where they began dismantling structures on deck to get wood for the cabin fire. They left the hull and the main decks intact and saw no signs that the ship was on the verge of breaking free anytime soon.

On March 11, the weather was sunny enough to use the astrolabe and take the height of the sun again. Twelve men were still able to stand upright and make the long trek for more wood. But doing so reduced them to a pitiful state, and they begged van Heemskerck for a cup of wine to dull their suffering. They knew nothing but the threat of death, which lay over them daily, could have motivated them to make the long trek outside. The sailors swore that they would have traded all their wages for more wood to burn, if there had been any way to buy it.

Just as they'd come to relish the sight of open water, ice drove in again, and the cold smothered them once more. A blizzard from the northeast dumped a sea of snow over their little hill, with thousands of flakes rustling like living creatures, frightening the men. The sight of the ice encasing everything once more drove them to despair, and temperatures plummeted as far as any they'd endured so far, making the sickest men even more ill. They spent a week pinned in the house growing weaker and weaker, with their wood reserves nearly consumed again. They had no idea what else to do. Daily fire was necessary to stay alive, yet they were too weak to get more wood. They made felt shoes from rough cloth and hoped for the weather to break.

The spring equinox came and went, but it grew no warmer. They'd saved the coal to take aboard with them if they were ever able to leave the island. But it would do dead men no good, they realized. The walls had turned to ice again, and the ceiling froze as well. Some of the crew argued that they should burn a little coal each day. On March 24, they couldn't go out. They closed up the house once more and put coal on to burn.

They made two trips for wood in fairer weather across the next six days and visited the ship. It lay empty when they investigated it, but it had been entirely taken over by bears, which had wreaked havoc on board. Two of the animals came near the cabin on March 30, but not close enough to cause the sailors trouble. After a time, the creatures wandered back toward the ship. In the distance, the men soon watched ice ramming and plowing over ice, raising huge hills where there had been open water only days before. On April 1, the weather was clear and cold, but they were too weak to trek out to gather harbor driftwood, and decided to burn more coal.

Barents took the height of the sun again, and the men once more tried to "stretch their joints" in a futile attempt to counter their scurvy. They improvised a long club and played a hockey-like game called *kolf*. While the wind changed directions, all the men healthy enough to walk visited the ship. There, they fed out a cable attached to the bower anchor to keep the ship from floating away if melting ice happened to free it in their absence.

By April 5, however, the run of fair weather ended. Far from breaking free, the ship wallowed in place, encased in more ice than before. The next night, the bad weather wasn't limited to wind or cold—the air was also wet with mist. A bear was spotted approaching the cabin, and the sailors inside went into their usual response mode, preparing their guns for the moment the animal moved into close range. As the creature bore down on them, they went to shoot. But no guns fired. Their gunpowder was no longer dry. Unaware of its reprieve from death, the bear continued down the rest of the stairs they'd cut in the snowdrift. It now stood entirely in view, coming toward the cabin door as van Heemskerck frantically tried to close it. A plank intended to bar the door sat above the frame, but in panic and fear, the captain held it shut with his body.

Faced with a closed door that was reluctant to open, the bear took its time. Eventually it turned around and went back up the stairs. Relieved, the men settled in for the evening. But two hours later, after night had fallen, they heard its footsteps approaching again. It walked a circle around the cabin, roaring. Trapped, Barents's men listened in terror. Without guns and in a weakened state, they'd be no match in hand-to-hand combat with a polar bear in daylight, let alone darkness. The animal climbed onto the roof and prowled overhead, striking the chimney until they thought the barrel would shatter. The bear slashed at their makeshift roof, with its planks and stretched sail. As they waited to see if the roof would hold or the bear would come roaring into their shelter, Barents and his men heard the thick canvas of the sail ripping above them.

But no hole appeared where the chimney stood. The roof

didn't give way. No polar bear fell from overhead, swiping claws at them. They endured the animal raging just feet away until, eventually, it left. During the foul weather that continued all the way into the next day, they listened for the bear's return with guns ready to shoot. Still, the creature didn't come back. In time they went up to survey the damage on the roof. The sail lay torn and pulled loose from its moorings along the chimney.

The weather remained vile over the next three days and into the fourth, but on April 8, they heard ice drifting out again with the current. Looking out from shore, they saw the sea once more. It resurrected the hope that before too long, they might go home. The foul weather seemed tolerable when it helped to clear ice from the coast, but winds and currents were fickle. April 10 brought all the ice back in again, and the days that followed piled it higher and higher, covering everything in jagged hills.

They fought despondency and dragged the sled for wood again on April 13. They'd made new felt shoes, and though they were tired, they delighted in their new footwear. The shoes served them much better in snow and ice than their wood or leather shoes—and now every man had his own pair.

The next day, they observed the ship from a distance, and saw more ice surrounding it than ever before. It seemed impossible that the ship hadn't already been crushed. They went to examine it at close range on April 15 and saw to their surprise that it still remained more or less intact. On the way back to the cabin, they caught sight of a bear stalking them, but they made motions with their pikes to defend themselves and the beast skittered away.

Walking the coastline where ice had shattered against ice, it seemed as if whole towns had been raised from underwater, with fully formed bulwarks and towers on their outskirts. They headed in the direction the bear had come from and found a hole in the ice.

It had a narrow entrance, which they approached. Thrusting their blades into the darkness, the crew found no living thing there. One of the men crawled into the opening but he couldn't bring himself to go far.

On their next visit to the ship, a group of sailors dared to do something they hadn't tried before. Climbing down from the ship, they walked along the hills and valleys of ice as near as they could to the open water. A small bird swam by. Seeing them, it dove to escape. The men reasoned that if the bird could dive, there must be even more water under the surface. They decided to take it as a good omen.

On April 18, eleven of the fifteen remaining crew members were healthy enough to join an expedition for wood. As they lay in their beds that night, they heard a bear on the roof once more. This time, they plotted offensive maneuvers, taking up weapons and going outside. The noise of their approach frightened the bear. The next day five men took turns in the barrel sauna.

They once again pulled their sled to the driftwood shore on April 20. This time it was loaded with a kettle and their clothing to save the effort of pulling even more heavy wood—not just for cooking but also for laundry—over uneven ice and snow all the way back to the cabin. They built a fire and boiled water along the beach, washing out the crew's shirts and drying them.

Several days of clear weather followed, including one that brought another bear to the house. They hit it with a shot to the body, sending it fleeing—all of which was observed by a second bear, which decided to give them a wide berth.

In the last days of the month, they took a reading of the sun again, played at ball and *kolf*, and visited the ship. During April's final hours, they looked up at the sky at night and saw the sun descend almost to the horizon then rise again. In the days and months that followed, the sun would always sit in the sky, without any darkness to add to their dread. They'd survived polar night.

It was no small accomplishment. Accounts of future expeditions during polar night would include murder, sailors drinking solvents, and medical staff overdosing on narcotics.[3] In 2018, one voyager socked in with a fellow traveler month after month during polar night at the Bellingshausen Russian research station would stab his companion in the chest.

The Dutchmen on Nova Zembla didn't have the luxury of narcotics to get them through months of darkness. But they'd nonetheless resisted any impulse to assault one another, and in winter's wake, May 1 turned out to be a good day. The midnight sun illuminated the sky, and the sailors cooked the dregs from their last barrel of salted beef. The barrel had been packed away in the hold of the ship the year before but remained edible. The sailors' only complaint about the beef was that it was now gone.

With rations thinning and the glory of daylight all around them, their minds began to turn to whether they might soon be able to leave Nova Zembla. Ice had fled the coastline again, but their ship still lay trapped in Ice Harbor. At some point, they would make their move, but the sea would have to coax the vessel closer to freedom to give them any real chance of sailing for home. The captain wanted to wait until the end of June—after the midnight sun had months to blaze its heat over the ice—to try to set out. In the hopes of keeping the crew strong for the work that preparing to leave would entail, van Heemskerck opened the last barrel of salted pork, and began dividing it among the men, with a two-ounce ration— less than a fat handful—given once a day to each man.

The ship had lolled on its side within a rock's throw of the open water since mid-March, but ice that had driven back in continued to accumulate until that distance had doubled and doubled again.

And still more ice charged in to block their freedom. The dream of freeing the ship wasn't quite lost to them yet, but the greater distance to the open sea meant less of a chance for liberation. Even if they surrendered the ship to Nova Zembla, the long expanse of ice also marked the distance they'd have to drag their two small boats up and down hills and ravines of ice, loaded with as many possessions and provisions as they dared to bring along. All this would have to be accomplished before they could even begin to sail. The weight of the load, the fear of damage to the small boats, and most of all the exhaustion of the scurvy-ridden men stood as depressing obstacles to departure.

Yet they longed to go. The sun began to drift higher in the sky in the evening, shunning the horizon. The sea pressed nearer to shore each day. But on May 7, a storm forced the sailors back into the house, where they grew restive and unhappy, fearing they'd never escape. The next day, they decided to talk to van Heemskerck and argue for departure from the harbor, which seemed it might never be free of ice. They knew they'd be arguing for the captain to move up his departure date by nearly two months and debated who could best present the audacious demand. But their plan faltered when they couldn't agree on a messenger.

Growing more unhappy by the day, the crew chose William Barents on May 9 to plead their case. Though he was already sick, he resisted their entreaties and calmed them. He heard them out without dismissing their fear. Not yet ready to mutiny, they let themselves be talked down from their demands. The following day proceeded like so many others: the sailors took the height of the sun, and the crew again surveyed the open water.

On May 11, the crew came to Barents once more to ask him to intercede with the captain. This time he said yes.

The men returned to everyday concerns. Days passed, bringing a snowstorm bracketed by afternoons in which the sea became ever more liquid. Still admiring their new felt shoes, they hiked out to the driftwood coast and hauled back a sled filled with wood. But on May 14, their patience began to wear thin. The men reminded Barents of his promise to talk to van Heemskerck.

The next day, the crew tried to restore their deteriorating bodies by walking, running, and playing *kolf*. Meanwhile, Barents shared the crew's wish for an immediate departure with the captain. Van Heemskerck agreed to leave before the end of June. But he didn't give them everything they wanted. They'd watch and wait for the two weeks remaining in May to see if the ship broke free or could be freed from the ice. If they could take their belongings back to a ship floating upright in the water, they'd set out as soon as possible. If not, they'd begin to modify their two smaller craft and make preparations to sail more than a thousand miles in open boats with no shelter from the elements.

CHAPTER NINE

Escape

For months, they'd endured in an in-between state—half dead and half alive—joining a long line of castaways whose deaths might pass unnoticed. Their former expedition mates and the ship they'd parted ways with back at Spitsbergen might be in the same situation or worse. But now the Nova Zemblan castaways could at least look toward their departure.

As they contemplated a return voyage, they knew they'd be going from misery to misery. Which castaways in prior centuries had found a way to return home on their own from uninhabited distant regions nearly a year after they were stranded? What they hoped to do seemed unprecedented. And yet they were happy that van Heemskerck was willing to leave earlier than planned, even as they fretted that the captain's compromise might still lead to unnecessary delay. The ship sat firmly in its block of ice, offering no hint of thaw. The small boats in their current state would never survive the kind of voyage the crew was contemplating—they'd take weeks of work to be ready for the trip. If they waited until the end of May to make a decision about abandoning the ship, they might end up stuck on Nova Zembla well into June.

The smaller boat of the two vessels would be particularly vulnerable. Wood and tools lay at hand to saw the boat in half and lengthen it, to help make it more seaworthy on any long voyage. But reengineering the boat was only the first challenge. Their crews would also have to somehow get it from the cabin to the sea. And dragging even a small boat to open water over the jagged hills and crevices in the ice would be a herculean task. Making the boat longer and heavier would only render the whole effort more grueling.

Three days of good weather set the men counting the hours until they'd sail for home. They visited the ship and hiked directly to the sea from the cabin, looking for the best path for hauling their boats into the water. The last barrel of salted pork was emptied about the same time the clear skies vanished on May 20.

As conditions packed ice back in along the coast, the crew spoke out once more, telling van Heemskerck at noon that if they hoped to leave at all, they'd better begin the hard work that would have to be finished before setting sail. Van Heemskerck answered that he valued his own life as dearly as any of them valued theirs, but the decision as to which craft they'd take would wait until the end of May. He urged them in the meantime to begin getting themselves ready and to take care of personal chores like patching clothes and repairing tools.

The next day they began to prepare in earnest. On May 22, with wood supplies low, they broke down part of the front porch and threw it on the fire. The next morning, they set out to boil water near the shore once more and do their laundry. By the twenty-fourth, enough ice had returned that very little of the sea lay visible. Barents again took the measure of the sun. Six days remained until the end of the month.

Under fair skies, the wind carried more ice in the next day, and on May 27, bitter weather did the same, heaping blocks and slabs in piles. Hearing doom colliding and accumulating in the harbor, the crew made their point more urgently, telling the captain it was past time to begin preparations. Van Heemskerck finally agreed.

With no way to free it, the ship that had carried them from Amsterdam now belonged to the sea. They'd have to take the things that they needed from it, and ready their two small boats to make a voyage that as far as they knew had only been navigated once in recorded history—by Barents, in a much larger vessel. The crew walked out to the ship and began poring over its treasures. They pulled down part of the rigging to take with them, and the old foresail as well. They'd need it to make sails for the small boats.

The morning of May 29, the sailors set out to find the scute, the single-sailed, flat-bottomed boat—the larger of the two they hoped to use. They planned to drag it to the house to start work on it. But heading to where it had been sitting since before winter set in, they didn't see it at first. The boat sat buried deep under months of snow. Digging it out exhausted them. Once it had been excavated, they stood looking at it and realized that they were too weak to move it to the house. The vision of the vast work that lay ahead of them if they were to have any hope of getting home struck them, and they lost heart.

The captain told them that if they wanted to go home, they'd have to rise up and do more than seemed possible. Otherwise, they could stay there as citizens of Nova Zembla and make their graves on the island. They were sorely disappointed by their failure, because they wanted to work, but found themselves unable to coax their bodies into complying. They hadn't caught a fox in more than three months; their scurvy was now advancing unchecked. Van Heemskerck and the sailors left the scute in the snow and trudged back to their shelter empty-handed.

Back at the cabin in the afternoon, they rallied a little, and decided to inspect the rowboat that already lay near the cabin. Once it had been turned right side up, they went to work and began building up the gunwales along the sides of the ship, to better fend off waves at sea. As they focused on their work, one sailor looked up and saw a polar bear coming at them.

They scurried to the cabin, where they took up their positions. The sailors with long guns moved to cover each of the three doors of the front porch, and a fourth shooter with a musket climbed up to man the chimney like a sniper. The bear moved toward the cabin more aggressively than any they'd seen before, heading right for the step at one of the doors. But the man armed with an arquebus at that entrance was turned toward another doorway and didn't see the polar bear approaching. It barreled closer, almost close enough to touch him.

His mates inside cried out in warning. The sailor spun toward the bear and swung the barrel of the gun up to shoot point blank. The weapon fired, its bullet passing through the creature and out the other side. The bear fled the scene but collapsed on the ground some distance from the cabin. The men inside ran with the rest of the guns and their half-pikes to finish off the animal. If the gun had misfired, as weapons of that type were prone to do, the bear would've seized their friend—and might have gotten into the cabin, too. Cutting the animal open to gut it, they found "remnants of seals, eaten whole" in its stomach.

The next day, the healthiest men went to work on the small rowboat outside the house, while the rest stayed inside making sails. But as the outside crew concentrated on modifying the boat, another bear approached. In a routine that had become familiar, they scurried inside once more, shooting at that bear, too. After it had gone, the sailors went on top of the house and began pulling apart the planking on the roof for use on their boat.

On the last day of May, the men returned to work, only to see a third bear headed toward them. Gerrit de Veer noted that the frequency of the visits made it seem as if the bears "had smelt that we would be gone, and that they desired first to tast a peece of some of us." Back into the Safe House they went, out came the guns, and three shots—two from doorways and one from the roof—took down the animal.

They hadn't enjoyed eating the meat from the first bear they'd killed on the voyage, almost a year before. But dwindling rations and the passage of time combined to make them look more keenly at this bear and reconsider. After gutting the animal, they dressed and cooked its liver, which had a much better flavor than the meat they'd eaten before.

They were pleased with their meal, but the bear had its revenge when the men started to feel ill. Everyone fell sick, and the cause was clear. Barents and his men had poisoned themselves. Polar bear liver contains enough vitamin A to be lethal to humans. Though the crew had no more idea of the effects of too much vitamin A than they did the lack of vitamin C that caused their scurvy, both wreaked havoc on the castaways' bodies just the same. Symptoms include drowsiness, headaches, liver damage, altered consciousness, and vomiting. The next morning, van Heemskerck picked up the pot of liver still sitting on the fire and threw its contents out in the snow. Three men soon lay near death.

No work was done on the boats that day, but the four men who were least sick from the poisoning made their way to the ship, to see what else they might gather for use on the voyage home. They came back with a barrel of salted fish, and each sailor got two.

Illness continued to plague them, but the sailors carried on with their plans. They investigated the best route from the boats to the open sea, deciding that despite the hilly landscape of the ice near the ship, the shorter distance still offered the most efficient path. By June 4—four days after they'd eaten the polar bear liver—most of the crew had recovered, but the skin of the three men who had fallen most violently ill peeled off in layers from head to toe.

Afflictions particular to their geography were a trial to them. Though scurvy ravaged many sailors, only in the polar regions were the means to defeat it so absent. Without frigid temperatures, they never would've had to resort to the coal fumes that nearly killed them the first time they burned it. And the polar bears that hunted them, as well as the animal's lethal liver, could be found only in far northern climates.

Yet their remote frigid location was in some ways a gift. Diseases like yellow fever, known to the Maya as "blood vomit," would curse many expeditions in North America, and affect villages and cities alike for centuries. But communicable diseases were hard-pressed to survive in the high Arctic. Diseases that felled sailors in other destinations around the world—from plague to malaria and smallpox—were nowhere to be seen in the far north.

More often, sailors would carry disease with them into new worlds. From smallpox to influenza and the plague, sailing ships were the vectors by which many illnesses made their way into other lands. In 1855 a ship carrying yellow fever docked at Hampton Roads, Virginia, killing thousands of residents. In the fourteenth century, rats from Crimea bearing the Black Death made their way onto ships that spread devastation across Europe. Even during Barents's era, smallpox traveled to the Americas easily under the brutal conditions of the transatlantic slave trade. But as far north as Barents and his men wintered, there was no one to give contagious diseases to but each other.

Though Barents, van Heemskerck, and the crew were spared tropical diseases, their northern challenges were agonizing enough. Fortunately, the three sickest men recovered from liver poisoning, leaving their shipmates doubly lucky. If three more sailors died before they could leave Nova Zembla, there might not be enough crewmen left to take on the tasks of preparing to sail or completing the approaching chore they most dreaded: hauling the boats to open water.

After six days of work, they finally got the rowboat into shape for the voyage. Then it became clear that they'd have to return to the scute that they'd been too weak to drag to the cabin before. Early in the morning, eleven men went down to the beach and began to move it. Whether the snow had become packed down and easier to walk on or the work on the rowboat had strengthened them, this time the scute was more cooperative. They dragged it toward the ship, where three men stayed to work on it.

It was a herring boat, and narrow at the stern. They lopped the back off and rebuilt it to be square, which would cut its speed in the water but make it steadier at sea. Just as they had with the rowboat, they began to build up the gunwales, to offer more protection from the waves. Neither craft was big enough to provide a cargo hold to let sailors hide from the elements. They braced themselves to sail for weeks, perhaps months, with no respite from the weather.

While three crew members worked on the scute, everyone else gathered food and equipment at the cabin. Loading up two sleds, they planned to bring the haul from the house back to the ship for storage, to more quickly provision the small boats once they were rebuilt and ready to sail. To their sorrow, the ship was in no danger of breaking free from the ice—it still lay half the distance between the cabin and open water. But it would work as a staging point to get the rations and gear partway along the path for departure—what was sure to be one of their hardest days yet. A waypoint would be a blessing and could also be secured against foxes and bears. As they worked, their hearts grew light, thinking that it might be possible after all "to get out of the wild, desart, irksome, fearefull, and cold country."

But almost a week of good weather ended with hail and snow, and June 5 trapped them indoors once more. They organized and gathered their masts and sails, their rudder for the stern, and the spar they'd use at the front of the ship for the bowsprit. They also packed rounded wooden flippers known as leeboards or whiskers that attached to the sides of the boats. The leeboards could be lowered or raised to keep a vessel from slipping sideways toward rocks or ice—a real danger, as close as they would have to stay to shore in their small boats.

The next day, better weather let the carpenters attend once more to the scute, while the rest of the crew hauled the sleds that had already been loaded. They carried with them food and some of the most valuable merchandise, which, with luck, they would return to investors back in the Netherlands. But by midday, the sky turned ugly again—bringing not only hail and snow, but rain as well. They'd already removed the planks from the roof to use on the boats. Now all that protected them from the elements overhead was a sail, which quickly began leaking under the weight of water in every form, liquid to solid. They were drenched. Their felt shoes, too, which had done so well for them on snow and ice, grew sodden from the slush that had taken over the path to the cabin. They set aside their improvised footwear and pulled out their old shoes, which had frozen so hard in winter.

They packed up more of the merchants' cargo on June 7, crafting tarpaulin covers for the salable goods, in the hopes of keeping them from the various forms of harm that might befall them in an open boat. The following day was clear enough for some of the men to take what they'd loaded over to the ship while the carpenters worked for a third day in a row on the scute.

Afterward, they gathered to try to bring the rowboat from the cabin to the ship and set it alongside the scute, which was now nearly finished. They ran ropes attached to the vessel over their shoulders and gripped them with both hands, letting their bodies do as much of the work as possible. A combination of hope and goodwill made the work enjoyable, and the men were pleased with their success.

On June 9, those who'd been drafted as carpenters finished the scute by laying the inside sheathing of the boat. The sailors not working on the scute took the opportunity to wash their shirts and linen by the shore one last time, with no idea if or when they would get a chance to do so again. The next day, they dragged four sleds of their belongings to the ship. The remaining wine went into small casks, to make it easier to divide between the two boats. They realized, too, that smaller casks would be simpler to hoist in and out of the boats onto the ice to leave them more room in the boat if they should get frozen in. They felt sure they would be frozen in, at least on the first leg of the trip.

On June 11, nasty weather prevailed once more, with a hard wind out of the north-northwest. As it roared around them, they sat inside the cabin and fretted that the ship would be blown away. They'd lose everything they needed to survive. But hours of agony spent shut in the cabin ended, and nothing was lost.

The following day, they hiked to the ship and pulled out their hatchets, halberds, and shovels. The next order of business was to cut a way through the jagged, frozen slopes that lay between the small boats and open water. They chopped out a rough corridor, throwing chunks of ice that could be lifted, pushing those that couldn't, and digging out the rest. As if to say goodbye, a bear appeared in the middle of their labor, rising out of the sea and stalking over the ice toward them. Only the ship's surgeon had a musket.

Gerrit de Veer sprinted toward the ship to get more guns, but succeeded only in drawing the attention of the bear, which turned to chase him. Polar bears can run faster than humans under most conditions, and after more than nine months on Nova Zembla, the men weren't in peak form. As the bear began to overtake de Veer, his shipmates came running behind and managed to distract the creature. When it turned to face them, the surgeon fired into its body, and the bear ran away, injured. Yet the hills and valleys of the harbor ice were so uneven that the bear couldn't easily go far. The men chased the animal down, and with their fear turning to fury, they smashed its teeth in its mouth where it lay and finished it off.

They had fair weather on June 13, with the carpenters finally finishing work on both the scute and rowboat. The wind was favorable, and the captain went down to look at the water, which seemed open enough to leave. After months of obstacles—including the twenty-three sightings of bears whose company had more than once nearly ended Barents's life, and the twenty-six foxes that had so far saved them—all that remained was to get the boats from the edge of the ice down into the water.

Van Heemskerck made his way back to the cabin and met with William Barents, who was by now too sick to help with physical tasks. The captain explained that conditions were good enough to set out. He ordered the crew "to take the boate and the scute downe to the water side, and in the name of God to begin our voiage to sail from Noua Zembla."

Barents had spent his time sick composing a letter, in case they all died sailing home. He'd written it out to leave in the cabin. Barents rolled the letter into a scroll and tucked it in a powder horn, which was capped and hung inside the chimney. Though their lodging might never be found, if someone should come across it by chance or by design, there would be a record of the fifteen Dutchmen who'd survived three seasons on Nova Zembla and tried to return home.

Van Heemskerck wrote a second letter describing their privations and their plan to set sail without their ship, leaving their fate in the hands of God:

Hauing till this day stayd for the time and opportunity, in hope to get our ship loose, and now have little or no hope thereof, for that it lyeth fast shut up and inclosed in the ice, and at the end of March and the beginning of April, the ice did so mightily gather together in great hills, that we pondered how to get our scute and boat into the water or where to find a conuenient place for it. And for that it seemed almost impossible to get the ship out of the ice, therefore I, with William Barents and the chief-boatswain and the other offi-cers and company of sailors thereunto belonging, considering with our selues which would be the best course for vs to saue our own lives and some wares belonging to the marchants, we could find no better means then to mend our boate and scute, and to prouide our selues as well as we could of all things necessairie, that being ready we might not loose or ouerslip any fit time and opportunity that God should send vs ; for that it was required for us to take the fittest time, otherwise we should surely haue perished with hunger and cold, which as yet is to be feared will go hard inough with vs, for that there are three or four of vs from whom in our work we have no help, and the best and strongest of us are so weake with the great cold and diseases that we haue so long time endured, that we haue but half a mans strength ; and it is to be feared that it will rather be worse then better, in regard of the long voiage that we haue in hand, and our bread wil not last vs longer then to the end of the mounth of August, and it may easily fal out, that the voiage being contrary and crosse vnto us, that before that time we shall not be able to get to any land, where we may procure any victuals or other prouisions for our selues, even if from this moment we did our best ; therefore we thought it our best course not to stay any longer here, for by nature we are bound to seeke our owne good and securities. And so we determined hereupon, and in general by us all subscribed, done, and concluded, vpon the first of June 1597. And while vpon the same day we were ready and had a west wind with an easy breeze and an indifferent open sea, we did in Gods name prepare our selues and entred into our voiage, the ship lying as fast as euer it did inclosed in the ice, notwithstanding that while we were making ready to be gon, we had great wind out of the west, north, and north-west, and yet find no alteration nor bettering in the weather, and therefore we have at length abandoned it.

Dragging the rowboat to open water along the path they'd cut in the ice, they left one man aboard to tend it. Those who were still healthy enough to work went back for the other boat. Filling eleven sleds, they traveled back and forth for food and wine again and again. In trunks and coffers, they loaded the most valuable of the merchant cargo: "six packs with the finest wollen cloth, a chest with linen, two packets with ueluet, two smal chests with mony, two coffers with the mens clothes such as shirts, and other things, thirteen barrels of bread, a barrel of sweet-milk cheese, a fletch of bacon, two runlets of oyle, six small runlets of wine, two runlets of vinegar, with other packs and clothes belonging to the sailors and many other things."

Once their treasures and hopes for staying alive were stowed, they took a sled back to the cabin and lay William Barents on it, hauling him down to the water. With Barents aboard, they returned for Claes Andries, who'd likewise been an invalid for some time, and lifted him into the other boat.

Van Heemskerck brought out the two copies of the letter he'd drafted. At the bottom it read "Dated upon the 13." He had each of the men sign it in turn.

> *Iacob Heemskerck.*
> *Willem Barentsz.*
> *Pieter Pietersz. Vos.*
> *Gerrit de Veer.*
> *Meester Hans Vos.*
> *Lenaert Hendricksz.*
> *Laurens Willemsz.*
> *Iacob Iansz. Schiedam.*
> *Pieter Cornelisz.*
> *Iacob Iansz. Sterrenburch.*
> *Ian Reyniersz.*

Four among the crew, including Claes Andries, didn't sign their names either because they were illiterate, or because they were too ill to do so. Van Heemskerck put a copy in each of the boats,

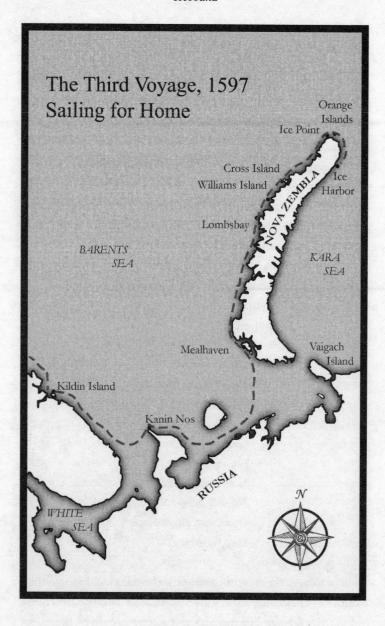

The Third Voyage, 1597
Sailing for Home

Orange
Islands

Ice Point

Cross Island
Williams Island

Ice
Harbor

NOVA ZEMBLA

Lombsbay

BARENTS
SEA

KARA
SEA

Mealhaven

Vaigach
Island

Kildin Island

Kanin Nos

RUSSIA

N

WHITE
SEA

in case they were separated by the elements or either vessel should perish at sea.

At half past four in the morning on June 14, 1597, they made their way to the edge of the ice. "And so," wrote Gerrit de Veer in his journal, "committing ourselues to the will and mercie of God, with a west north-west wind and an endifferent open water, we set saile and put to sea."

The renovated boats passed their first, most basic test, and stayed afloat. As the wind drew them away from Ice Harbor, their former ship became indistinct in the distance. Left behind, it would thaw and refreeze, unraveling in the slow deterioration endemic to the Arctic, the cargo hold filling with more and more water, until the inside and the outside of the ship were fully of the same medium. The objects they left behind—the cannons, the cargo that had been left belowdeck, the planks of the quarterdeck, the three tall masts and the rigging with its halyards and sails—would sink to the bottom of the harbor or drift out to sea with the currents, disintegrating in such slow motion that an observer might have to watch a week, a year, or a century to observe changes. The ship would remain untouched by humans for four hundred years.

Their cabin would sit up on its low hill with a view of the ship's demise, and would itself go undiscovered for three of those centuries, filled with stacks of cheap prints, pewter candlesticks, a frozen clock that stood knee-high on a Dutchman, and William Barents's discarded book about the landscapes and history of China.

With the aid of a westerly wind, Barents and company sailed northeast from Ice Harbor. They made it out of the harbor, but their biggest fear quickly came to pass. The boats got pinned by ice, and they couldn't work themselves free. Four crewmen went ashore and climbed to higher ground to get a clearer view. Along the cliffs, they lucked into four birds they killed with stones, the first fresh food they'd eaten in weeks.

On June 15, the ice moved away, and they set out once more. They covered fifty-two miles, following the jagged coast north to Cape Desire. The next day, they sailed another thirty-two miles, making it to the tiny Orange Islands off the northeastern coast of Nova Zembla. Sailors hauled barrels and a kettle ashore, to melt snow and ice for drinking water to carry with them out to sea. They built a fire and looked for more birds or eggs to cook for the sickest men, but they came up empty-handed.

Van Heemskerck and de Veer had better luck, going with a third sailor across the ice to the smaller of the two islands. Jagged paths led up to the top of steep cliffs, where they could edge their way within arm's reach of any number of birds nesting in stone crevices. With the birds unaccustomed to any human presence, the sailors managed to grab three. But while they were carrying their prey back across the short distance between the two islands, van Heemskerck fell through the ice. A strong current underneath the frozen surface dragged him away, but he managed to pull himself out of the frigid water and claw his way back, where he sat by the fire until he was dry. Meanwhile, the birds were dressed and fed to the men most afflicted by scurvy. Like the foxes, the birds' flesh contained vitamin C. If it wasn't enough to rid them of scurvy, at least it might let them live a little longer.

After melting sixteen gallons of fresh water and loading it into the boats, the crew set out again in the "drowsie miseling weather." The boats offered no cover to the men, and everyone was damp or wet. They headed toward Ice Point, which stood out in relief on the map they'd made. Once both vessels had arrived, they drew up against each other to allow clear communication. Van Heemskerck called out to Barents in the other boat to see how he was feeling.

Barents called back, "Quite well, mate. I still hope to be able to walk before we get to Wardhuys."

They all knew that passing Ice Point meant they were rounding northern Nova Zembla and could soon head south, where temperatures would begin to rise and ice might plague them less. Barents had de Veer in the boat with him. The navigator turned from talking to van Heemskerck and asked, "Gerrit, if we are near the Ice Point, just lift me up again. I must see that Point once more."

They stopped soon afterward, fastening the boats to drift ice so they could eat. As they had their meal, the weather only grew more miserable, and soon they were pinned in again by floating ice. They spent the night under the midnight sun, but morning only brought more danger piling in around them. It had been frightening enough to be at sea amid colliding icebergs in the ship a year ago at the same time. Sitting in the fragile, makeshift boats as the current drew them pell-mell among frozen blocks was terrifying. They felt the ship would be smashed to pieces any moment, and realized that if they couldn't find a way to secure the boats away from the treacherous moving landscape, they'd die. As they were swept away, they looked at one another in despair.

Someone suggested that if they could only secure a rope or tackle on the fast ice—ice attached to the coast or sitting on land below the water's surface—they might be able to pull the boats to safety. But far from shore and caught in the current, how could they manage it? One of them would have to leave the boat and carry a rope toward land across the moving lanes of ice—all shapes and sizes, many of them slippery. Anyone who dared to go out would have to climb from one floating platform to the next without knowing how it might shift or whether it could support the weight of a man.

Gerrit de Veer believed that whoever tried to reach the fast ice could easily end up carried away on the frigid current. But if someone didn't try, they'd surely all perish. Knowing he was the lightest of the sailors, de Veer seized a rope and climbed out of the boat. Stepping onto a piece of moving ice, he worked his way toward the shore, creeping from block to block, trailing the cable behind him. Coming to a frozen ledge that was anchored to land, he looped the rope over a hill and tied it off. If de Veer had tried to pull the craft filled with sailors to shore on his own, he would've made little progress, but a secure line made all the difference. Sailors on the other end began to haul, slowly dragging themselves toward safety.

Once both boats had been brought alongside the ice, those sailors still strong enough to carry their mates lifted Barents and the other invalids out onto the frozen surface, laying clothing under them to keep them comfortable and as warm as possible. After the sick men came the provisions, hoisted in haste before the boats could be smashed to pieces. Once the boats sat empty, the crew dragged them up from the water, too. Staying there all that day, the men made repairs to the craft, which had been stoved and battered by the ice. Enough extra wood could be scavenged to build a fire, which let them melt pitch to repair seams on the boats. They laid tarpaulin for waterproofing.

Once the boats were seaworthy again, some of the sailors went ashore to hunt eggs for their sick mates. They fell and got wet on the hunt, nearly losing their lives as they slipped between ice and land. No eggs could be found, but they returned with four birds. The next day, with ice surrounding their ledge, they saw no way to escape. After abandoning the shelter of their cabin, which had so often felt insufficient to protect them from the cold, they now had to huddle in their boats atop the ice. They came to think that they might die there, but made light of the possibility. Consoling one another, they recalled the other scrapes they'd escaped through what felt like divine intervention.

On the morning of June 20, the ailing Claes Andries declined further. The sailors from his boat realized he was dying. His nephew John was also on the voyage, and also ailing, but wasn't as close to slipping away. The chief boatswain came over to William Barents and the sailors in the other boat to tell them that Andries didn't have long to live.

After the boatswain made his announcement, Barents spoke up, saying, "Methinks for me, too, it will not last long."

The comment took de Veer and the other sailors by surprise, because they hadn't thought Barents lay near death. De Veer spoke with his mate at length. He showed Barents the chart he'd made of their voyage so far. Barents discussed the map with him, until finally he put it aside. "Gerrit," he said, "give me something to drink."

Barents drank what he was offered, but just as he finished, his body quivered where it lay. His eyes rolled up in his head, and he lost consciousness. His pulse soon vanished. There wasn't even time to call for van Heemskerck to come from the other boat and exchange last words with his navigator. William Barents was dead.

Claes Andries managed to outlive Barents, but died soon afterward. Yet it was Barents's name that would become immortal.

Even during his life, Barents had lived a larger life than most humans. He'd been the first to publish an atlas of the Mediterranean, a survivor of nearly ten months in some of the most extreme conditions on the planet, a three-time explorer into the unknown, mapping places no European—and in some cases, perhaps no human—had ever seen. In Barents's day, the Russians called the sea between Scandinavia and Nova Zembla the sea of Murmans, referring to the Norwegians they encountered there. But in 1853,

Barents's name would come to replace the earlier one, and the waters he sailed three times on his way east would come to be known worldwide as the Barents Sea. Four hundred years later its treacherous conditions would lead some to call it the devil's dance floor.[1]

In time, the machinery of commemoration would take fierce hold of Barents's name, and the Dutch would embrace him as a national hero. He'd be transformed into an icon whose actual achievements and heroics were warped into a tribute to the greatness of empire. It wasn't enough to have been a skilled seaman, a scientific observer, and a committed explorer who held steady in the face of terrifying conditions. The sparse biographical details left to history would be taken up, filled in, and transformed. The possibility that Barents had come from the island of Terschelling would, for many, morph from likelihood into a clear statement of his origins. A backstory of his life—that he was the son of a farmer, a veteran at the age of twenty of a naval battle on the Zuiderzee—would be invented when verifiable information couldn't be found.[2] Two towns would claim to be his birthplace.

Before Barents, other Europeans—Columbus, Magellan, and Vasco da Gama—had gained renown for their explorations, either because they helped to find new continents unknown to Europeans, or a sea route around them. Their exploits had brought tangible assets: gold, silver, or trade and a stake in habitable new lands.

Barents, too, mapped new lands, but his legend took on a different form. Though he'd sailed farther north than any European on record, he failed to find an open sea route. Others would continue the quest for a northeastern or northwestern passage, with Swedish-Finnish explorer Nils Adolf Erik Nordenskiöld complet-

ing the former in 1879 and Norwegian Roald Amundsen the latter in 1906.

Yet Barents's name would come to possess surprising staying
power. After Barents, the polar regions weren't simply a byway to
the East, but a new frontier in and of themselves, a force to be reckoned with, an obstacle.

This shift happened in part because while Barents and his fellow
castaways tried to sail home from Nova Zembla, other Dutch sailors
were making their way back to the Netherlands as well. The remnants of the fleet that had set out to find a southern route during
Barents's second voyage two years before were headed home and
would arrive that August. Four ships carrying two hundred forty-
eight men had rounded Africa then continued on to the East Indies.
The maps prepared by Petrus Plancius had carried the fleet as far as
Java and Bali in the South Pacific—halfway around the world.

Petrus Plancius's charts, based on Portuguese and other intelligence reports, had worked well enough to get the ships to their
destination. But carnage and incompetence had dogged every
other part of the voyage. Scurvy ran rampant in the first months
of the expedition. Widespread contagious disease led to the delib-
erate burning of one of the ships to forestall a fleet-wide epidemic.
The eventual commander of the fleet, Cornelis de Houtman, had
botched negotiations with a sultan who'd hoped to trade with the
Dutch, and was met with resistance and occasional open hostility.
Misrepresenting the approach of an indigenous royal family on its
way to greet him, de Houtman opened cannon fire on them. And
perhaps unsurprisingly, given the events of the voyage, he inspired
a revolt against his leadership. The fleet returned missing one ship
and carrying only eighty-seven men and a tiny cargo of spices and
black peppercorns. Nearly two thirds of the crew had died of illness or been killed in fighting.

Outside of the shared, and nearly universal, experience of scurvy on the high seas, Barents's final voyage stood largely in opposition to that of the southern expedition. The Nova Zemblan crew had set out to deliberately sail into unknown waters, rather than to follow a trade route established by prior explorers. On their mission, they hadn't murdered native populations or mutinied in the face of inhuman conditions. Though Barents and the captain of the other ship on the expedition had argued and gone their separate ways, their disagreement didn't end in a revolt between factions. Yet Barents, van Heemskerck, and their crew had failed at the one thing that the other expedition had accomplished: securing the Dutch a navigable route to the Pacific.

The expedition to the East Indies, however miserable its execution, did establish viable trading partners that could be coaxed or bullied away from the Portuguese. For this reason, along with its small but lucrative haul of spices, it was hailed as a success. A second voyage would quickly be planned.

With the East Indies expedition capturing the future of global trade for the young Dutch nation, William Barents's futile effort in the north would transform into something else altogether. Barents's expeditions and death launched another identity for explorers: the beleaguered polar hero. Rather than successfully connecting one part of the habitable world to another, these heroic explorers would find their legends bound up with unfathomable suffering and endurance.

The alternate version of history—in which a trade route could be found that didn't originate in the massacre of native populations or the African waypoints that would in time bring the Dutch into the slave trade—vanished into the ether. After the science and mapping that Barents and the crew had conducted on the voyage, all that was left was cold, suffering, and their attempts to survive.

Before Barents ever set out, Pomor sailors in Russia, as well as Englishmen Hugh Willoughby and Martin Frobisher, had each searched for a northern route between the Atlantic and Pacific Oceans. Arthur Pet and Charles Jackman had sailed all the way through the strait at Vaigach and had, like Barents, seen the Kara Sea. But by heading due north as far as he could sail at the beginning of his third expedition, Barents made—as nineteenth-century Dutch explorer L. R. Koolemans Beynen called it—"the first true polar voyage." The fact that the other ship, helmed by Jan Cornelis Rijp, tried to follow that mission more devotedly by continuing northward after Barents split the fleet, or that Barents ended up overwintering on Nova Zembla hundreds of miles from the Pole, is simply emblematic of his larger story. Barents's failures came from the immense adversity he faced. In time, that adversity itself would become the source of his fame.

With his time in the high Arctic and his overwintering, however inadvertent it had been, Barents would become the first face of the many polar explorers who followed in his wake. After Barents would come Henry Hudson, who in 1608 would fail to get even as far as Barents in his effort to sail over and beyond Nova Zembla. Nearly three centuries later, a series of expeditions would try to reach Barents's cabin, only to fail.

A melodramatic and wildly inaccurate 1819 poem about the overwintering written by Hendrik Tollens—who also wrote the Dutch national anthem of the era—recast the story in telling ways. In the jingoistic poem, it's Jacob van Heemskerck who decides to set out, looking for another challenge once the Dutch had already established a southern trade route to the Indies. The two ships on the voyage are driven apart by a storm, rather than the navigation dispute that actually split the expedition. Once on Nova Zembla for the winter, the men sleep in the snow in darkness while a bear sneaks up on them and drags away one of the sailors, taking him back to its den to eat. The crew wakes up but has no idea what's happened. Only when they do a roll call of names do they realize one of them is missing. In the morning they see the bloody trail left by their doomed friend. After they build their hut, the men are besieged by several polar bears at once. The poem would lionize Barents and etch his reputation into the heart of every Dutch schoolchild and many foreigners as well, but it would do so at the cost of accuracy.

The century in which Barents would become larger than life—the nineteenth century—would also turn out to be the era in which the modern mythology of polar explorers crystallized. As technology delivered newspapers, telegraph wires, and later radio, each medium made it possible to track the progress of explorers heading farther and farther north, until they eventually flew over, landed on, and trudged to the North Pole itself.

But during the nineteenth century, expeditions would still go awry in horrible ways. Technology and money were no guarantee of success. After the horrors of cannibalism on the Franklin expedition, historian Beau Riffenburgh explains, the love affair with nature, and the romance of sailing into the unknown was replaced by the obligation to subjugate nature by "filling in blank spaces on the map." It's no wonder that Barents's reputation grew during this era: his suffering and death could be warped to fit cleanly inside this new view of Arctic exploration as man's struggle against nature and his attempt to dominate it.

Meanwhile, as the end of the century approached, a cult of suffering emerged. Circulation wars between newspapers promoting various explorers and expeditions dramatized stiff-upper-lip accounts in which audiences could agonize along with their heroes almost in real time. These newspapers would play a key role, but just as the book narrating Pet and Jackman's expedition had crossed Europe in his day, books would remain the chief vehicle by which the celebrity of explorers would be established.

Books paired with worldwide lecture tours for explorers would become the expectation, though not everyone who aspired to that status managed to achieve it. Fridtjof Nansen, Robert Scott, Robert Peary, and Roald Amundsen all became public heroes—their physical prowess and fearlessness transforming them into minor gods. But men like Umberto Nobile, who designed and piloted the airship *Norge* to sail over the North Pole, or Hjalmar Johansen, who went to sea on polar expeditions with Nansen and Amundsen, would find their contributions diminished or ignored. Even those who achieved fame and ran successful expeditions still often struggled to attract patrons.

Barents, of course, had failed to deliver on his Arctic mission in nearly every way. He'd lived through one near-mutiny and witnessed another full-fledged revolt before succumbing to death far from home—not fighting a bear, not poisoned by his crew, but on the ice still in view of the land whose hardships he'd defied for nearly a year, but in the end, couldn't escape.

Still, there was something larger than life in Barents's living long enough to see all of his epic plans fail. He'd staked the ship and his life on his pursuit of a northern route, and he'd lost everything. He was the patron saint of devoted error, living the consequences of his mistakes. He found no northern route and no path to China. He wasn't even fully in charge of the expedition. Yet it was to him the men came when they desperately wanted to convince the captain to abandon the Safe House and head home. If van Heemskerck had become their knight, the one they sparred with over leaving the island, the one who'd stood with Gerrit de Veer against three bears at once to save his crew, the one who held the door of the porch against another angry beast, Barents had been their magician. He could look into the sun and fix their ship's position on the globe. He could watch the stars and tell them the day of the year. He knew when the sun was in its fixed place and when time had slipped out of kilter.

In the end, William Barents managed to survive winter on Nova Zembla, but died nestled on the ice, somewhere between the islands and the sea, claimed by both but found in neither. His grave would be sought, though there is no record of the men digging one or laying stones over his body. Russian explorers Dmitri Kravchenko and Pyotr Boyarsky would link historical references to contemporary locations or identify stone cairns and markers along the Nova Zemblan coastline that might have covered Barents's remains—including one with a nearby polar bear skull sporting a gaping hole similar to that made by a bullet.[3]

Meanwhile, Dutch maritime historian Diederick Wildeman suspects that the men wouldn't have expended effort on burial in their desperate, weakened state. In all likelihood, there wasn't enough time left to them—or strength in them—to bury William Barents. With a choice between Nova Zembla and the sea, perhaps his shipmates left him on the ice, letting him finally flee the land that he couldn't escape in life.

His legend would grow into something unrecognizable, but what he actually accomplished was astounding enough, both human and extraordinary. Leaving Barents's body at sea or on shore, captive forever to the high Arctic that would never again be closed to a European presence, van Heemskerck and the rest would have to try to carry on without the man they'd believed in, the one whom Gerrit de Veer had called "our chief guide and onely pilot on whom we reposed ourselves. But we could not strive against God."

CHAPTER TEN

Staggering Homeward

The surviving thirteen crew members spent another full day
stuck on their frozen ledge before the sea cleared enough for
them to venture out again. They now had been gone more than a
full year. They were already presumed dead; the world had moved
on in their absence.

They had lost their chief navigator and, with him, some of the
confidence in their ability to carry out their final, desperate enterprise. But the celestial navigation at which Barents was so skilled
wouldn't be the kind they'd rely on for their homeward route.

Even while Barents was living, the sailors hadn't planned on returning the way they'd come—crossing the open sea along the 75th parallel that had already brought them from Spitsbergen to the eastern coast of Nova Zembla. They would have to supplement their remaining provisions and get fresh water, which would require them to make landings along the way. Even if they had enough food and water to sustain them, their small boats already desperately needed repairs, after only a week of sailing. They'd surely require more. But even if no repairs were needed, their boats were too small to take on the open sea. In high waves, the sea could swallow them whole.

Instead, they planned to work their way hundreds of miles south down the entirety of the coast of Nova Zembla to Vaigach Island—the site of the doomed second expedition. From there, they could follow the Russian shoreline and nearby islands westward along the continent, as they curved north to Lapland then Norway. Nearly all this coast had been mapped during their first two voyages, and de Veer had the chart from the third trip. If they stayed in sight of land, they wouldn't lose their way, and could hunt for safe harbors at which to stop along the route.

On the morning of June 22, they had seen enough open water to escape. But it didn't reach their campsite atop the iceberg. They'd already learned the answer to this cruel puzzle at Ice Harbor. If the water wouldn't come to them, they'd have to go to the water. Once more they lowered their boats over the side of the ledge, filling them with invalid sailors and provisions, and red cloth, and the rest of the most valuable cargo that they had so far preserved.

Arriving at the next iceberg between them and the open sea, they climbed up and hoisted the vessels after them. Atop the second iceberg, they hauled their boats and all the provisions another hundred feet before lowering their craft into water from which they could set sail. They were haggard and tired and knew that death was counting the days until it could claim them, but they kept moving.

Finally back in open water, they sailed both west and south before ice found them again. Once more they were trapped. But without warning, the ice gaped open like a lock on a canal, and they rushed in—only to watch the ice return. They took advantage of their captive state to eat, but having eaten, they remained stuck. Hacking and striking the ice did no good; they were forced to wait for the current's indifferent gods to open another way for them.

The next day, they made it as far as Cape Comfort, a hundred miles along the coast of Nova Zembla, before ice blocked their way again. They used their mariner's astrolabe and recorded the height of the sun to update their chart, but their minds were elsewhere. Though ice was everywhere, their supply of fresh water had almost vanished. They laid plates of snow out in the boats, to let the sun's heat melt it. They put handfuls of snow in their mouths to liquefy it even faster. But for the first time on the voyage, a dreadful thirst began to stalk them.

The desire for open water and a steady wind was strong, but their range of acceptable courses grew narrower. They were used to navigating the dangers of a lee shore, where a strong wind that blew them toward land and onto rocks or ice could smash their boats. But now a fierce wind that pushed them too far from shore became just as dangerous. Though it might kill them more slowly, heading too far out on the open sea would invite waves to swamp the boats as they drifted farther and farther from any source of fresh water. They had no choice but to make slower progress, pick their way through the ice, and follow the shoreline as much as they dared.

The morning of June 24 found them still looking for openings in the high ice. They got their oars out and tried to row their way through the maze towering around them, though they could see no open path. When an opening appeared later in the day, they sailed into it, trying to get around an outcropping of land, but drift ice again blocked their way. They decided to head for shore. Six men set out on the beach to gather kindling and look for birds or eggs, returning only with some wood. Once aboard again, they boiled a pot of crushed biscuit mixed with snow to make a porridge they called *matsammore*, which at least put something warm in their stomachs.

Meanwhile, the wind rose higher, and they secured the boats to fast ice to keep from being blown into the open sea. But the fast ice broke loose, dragging their craft farther from shore, and the sailors had to work to free themselves before they were lost. They bound themselves to another piece of fast ice on June 26, but the south wind continued relentlessly for another day, likewise cracking open the new ice they were using to anchor the boats. The crew couldn't move fast enough to stop what followed. Without warning, the rowboat and the scute were both dragged into the current, separated, and driven out to sea.

In Gerrit de Veer's boat, the men brought out the oars, and tried to row their way back toward the coast. But they couldn't row against the current with enough strength to make real progress. Reluctantly, because of the high wind, they hoisted their foresail, hoping to use its power to achieve what oars couldn't. But the slender foremast shattered, and then snapped in a second place, a devastating blow to the tiny boat.

They couldn't let themselves be driven out to sea, where they'd surely die of thirst if they didn't capsize first. Though the wind still blew just as lethally, they hoisted their mainsail, which—once it caught wind—they knew had just as much risk of splintering their mainmast. Water quickly surged over the gunwale, and the sea emptied out, until they were tilting far to one side. Looking down into the abyss, they saw "nothing but death" below them. They took the sail in before they could be drowned along with the boat and waited to see what would follow.

Suddenly, the wind changed direction, and they made their way with caution and fear back to the fast ice at the shoreline. As the vessel moved into calmer water, they began to hunt for the other boat and its crew. Sailing four miles along the coast, they looked and listened, but found nothing. Fog and mist obscured their view. They were filled with dread that half their company was gone. One sailor thought to pull out a musket and load it. A shot rang out, and after it, silence.

Then an answering shot came—not from anywhere visible, but a clear reply. Somewhere not too far off, sailors in the other boat were still alive. Continuing on their course, they found their shipmates' boat grounded between driving ice and fast ice, unable to move. Climbing out onto the fast ice that lay between them and the other boat, they worked together to unload provisions from the scute and drag it across the ice to open water again, reuniting the vessels. Some sailors had already fetched wood, and when both boats were afloat again, bread and water were once more boiled up and served hot.

The wind filled their sails the next day and blew them past the Cape of Nassau, which they'd hoped to get to days earlier. Then the air and sky turned against them. They struck their sails and pulled out the oars, settling for the less elegant, more painful progress of rowing. Never straying too far from the fast ice along the coast, they spotted piles of walruses spread over the ice. Even better, they found birds, shooting and retrieving a dozen before fog socked them in. Soon after losing visibility, they began to slip back into moving ice and had to secure themselves again to wait out the bad weather.

On June 28, icebergs pressed in harder and once more left the sailors expecting that their boats would be ground to bits beneath them. Knowing what had to be done was surely no comfort, because it would be agonizing to do it. Again, they unloaded their cargo and provisions. Again, they hoisted the boats up onto the ice. Spreading sails over the boats like tents, they watched more ice gather all around them. After setting one man to keep watch, they lay down to sleep.

The midnight sun still reigned, hiding the stars. As the sun moved into the north, the sleeping men, tucked inside the boats under tented sails, heard the watchman cry out. "Three bears! Three bears!" The sailors scrambled for their weapons, which were not at all loaded for bear, but were instead filled with shot for the birds they'd been hunting along the shore the day before. Spraying shot into the huge creatures, they spooked their attackers into retreat.

As the bears lumbered away, the men loaded bullets into their muskets and fired at the fleeing animals. One was killed outright, and the other two ran until they could no longer be seen. But two hours later, the surviving creatures circled back. The men made such a racket the animals were once again frightened off. It was no pleasant thought to realize they were out on the ice, in the bears' very element, with no cabin or wall to protect them. Meanwhile, the ice piled up in layers, a frigid horizon trapping them in place.

The bears were the first they'd seen since setting out in their boat two weeks earlier. The next day, despite the noise and bullets the men had unleashed the evening before, the creatures returned to eat their fallen comrade. After chasing off the bears, the men wanted to put the carcass of the dead animal in a high spot visible from their boats, to give them advance warning of any more visits. They wondered at the strength of the beasts, having just seen one creature carry its dead companion "as lightely in her mouth as if it had beene nothing," while four bedraggled sailors wrestled to move the half-consumed animal with difficulty.

On June 30, the men looked out to sea and saw two bears riding a piece of ice toward them, as if planning to attack. But the animals balked. The men came to suspect that they were the same two bears. After the pair retreated without violence, another bear appeared on the fast ice by the shore and made a beeline toward the boats. But it was easily scared away by their loud response, leaving the humans to watch and worry in the mist and wind.

The next morning, that bear or another like it climbed down from the currents of ice into the water and swam toward them. It made its way onto their ledge, but again, they frightened it away. The ice had opened enough for the polar bear to swim, and perhaps, they thought, even enough for them to put out to sea. But later in the day, the drift ice drove in toward the fast ice, ending all thought of departure. Incoming icebergs collided with their ledge, shattering it. The ice began to break apart beneath their feet. Frozen blocks tumbled pell-mell with the crew's belongings, as the sailors and boats fell, too. Provisions and cargo alike dropped into the freezing water.

They first moved to save the rowboat, trying to climb up the remaining ledge to haul it closer to land and out of danger. As other men began to save the food and cargo, they grabbed one treasured item, only to watch others fall into the water. Those who went to haul the scute up toward the rowboat found only more trouble, as the ice gave way under them again, sending both sailors and boat into a swift current filled with icebergs. Those who tried to gather the scute's provisions similarly lost their footing. As they staggered over the uneven surface, ice broke beneath them again and again.

Though it was the larger of their boats, the scute was battered as mercilessly as the men. It broke open at the seams and places they'd rebuilt for the voyage: the mast, the supports, and the corner of the boat where a sick sailor and a chest of money had been nestled. As they went to retrieve the latter, the sheet of ice they stood on shot sideways, slipping under another ledge that drove into it like a wedge. The boat vanished from view under a layer of ice.

The crew members looked helplessly at one another. They couldn't all fit in the rowboat, and there wasn't sufficient time or wood along the coast to build a new vessel. Without their scute, they were lost.

Without warning, the sheets of ice drove apart again, and they caught sight of the scute once more. Before the terrain could shift a third time, they ran to the scute and began dragging its damaged frame over to where the rowboat had already been secured. The violence done to it had wrecked it. The rest of the day was entirely consumed at hard labor, making the beginnings of repairs to help it float again.

It was the worst day of the voyage so far—worse even than their shock and grief over the loss of William Barents. They'd saved the boats, but they'd nearly drowned in doing so. And the sea had claimed from them a trunk of sailors' clothes, a chest of linens, a packet of scarlet velvet, navigation equipment, oil, cheese, and two precious barrels of bread. Almost as dispiriting was the loss of a cask of wine that, smashed open, had bled the entirety of its contents onto the ice without a drop saved. They sat exhausted, cold, and frightened in the face of what they'd already survived, unable to contemplate the treacherous canyons of ice scattered over the jagged white miles that still lay between them and home.

The morning of July 2—a year to the day since Barents had split with Jan Cornelis Rijp's boat to sail east toward Nova Zembla—began inauspiciously, with the sight of yet another bear stalking them. They managed to frighten it off without a battle, then set to work on the scute again. Six men pulled out some of the bottom boards lining the inside of the boat to use for repairs, while six more went to look for driftwood and stones to build a fire. The stones would keep the wood from getting wet, and a fire would let them melt pitch to waterproof their repairs. They also hoped to discover logs substantial enough to replace the broken mast.

When they returned they were carrying both stones and wood, and some of the wood they found had been worked with an ax. The signs of human work seemed freighted with meaning. Barents and his men had set out for China, hoping to find a new route to an ancient civilization. Now they'd become archaeologists unearthing relics, trying to find a way back to their own world in the present. Though the carved wood might simply have washed ashore from the mainland with the driftwood, it was nonetheless a sign that they'd begun to return to the land of the living.

Building a fire, the crew heated their pitch and went back to work on the boat. After they finished, they boiled the birds they'd previously shot and ate well for the first time in two weeks.

The next day, two crewmen had the strength to go exploring in the bitterly cold water. They recovered a pair of oars and a rudder, as well as the packet of cloth and the chest of linens. It was too much for them to drag back, but they carried what they could, bringing along a hat that had been packed in a trunk, which probably meant that ice had smashed the trunk open. On hearing the news, van Heemskerck took five more men back to the site, where they dredged everything they could find out of the water. The packet of cloth and the chest of linen were flooded and waterlogged, making them too heavy even for the group to carry for long. But the sailors left them out on the ice to drain, thinking they would come back for them before setting sail again.

When the crew next lay down to sleep, a bear made its way to their outpost. One man stood watch but didn't notice the animal as it moved toward them. The creature was nearly close enough to seize him when another shipmate caught sight of it and called to him to beware. He fled while a third sailor shot the bear in the body, frightening it away.

July 4 marked the most glorious day of weather that they'd seen in their time on Nova Zembla. They melted snow and took the bolts of red cloth that had been soaked in saltwater, rinsed them clean in fresh water, then dried them. The luxurious fabric could do them no good on the trip, but their efforts to preserve it would show the expedition's investors that the explorers had done everything possible to preserve the most valuable cargo they'd carried aboard.

On the next day, John of Harlem, the nephew of Claes Andries, lay on the ice and breathed his last breath. The loss of a fifth man, the third to die in three weeks at sea—reduced the company to twelve. If he wasn't buried on land, he was likely left on a floe or lowered into the water with a prayer.

The ice paid no respects to the dead and continued to appear, stalling their progress another day. When it was clear they couldn't leave, six men went to shore to find firewood to cook their meat. On July 6, the day began in mist and fog, and the adjacent sea remained locked to them. But near evening, the skies began to clear.

Some of the sailors walked to open water the following day carrying guns loaded with shot and killed thirteen birds. Riding drift ice to pick up the creatures, the hunters brought them back to the fast ice to dress. Though the weather turned nasty on July 8, they managed to cook the birds and sat on their frozen ledge eating their princely feast.

On July 9, the ice began to move out on the current again, making an opening on the shore side of their ice ledge. The captain took some men to reclaim the packet and the trunk they'd left out to drain, and the sailors packed them into the scute. But to keep from being mired in another collapse, they'd pulled the boats far from the edge of their iceberg, and the crew had to drag the boats a thousand feet or more just to get to a point where they could be lowered into the water. A year into scurvy with only twelve men remaining out of the original seventeen who'd gone ashore on Nova Zembla, the haul was excruciating.

They eventually set out with an east wind, but they saw the route wasn't yet clear enough to sail. Late in the day, they were forced to turn back to the fast ice. Another attempt followed the next morning, as they cautiously rowed their way through a maze of moving blocks. But once the original danger was past, another obstacle arose. They emerged to see two large plains of ice directly ahead of them come together, watching in disappointment as the path between them closed. The icebergs were too vast to row around, and they had no choice but to climb again onto the ice and draw their boats up. Once up, they dragged the vessels two hundred fifty feet or more to cross one plain and get back into the water.

After they descended once more into the sea, the boats slipped between two new moving fields of ice. As they followed the course between them, the icebergs began to converge. The gap between the boat and the wall of ice on each side grew smaller and smaller. They realized they would soon be crushed. Hoping to outrun death, they pulled out the oars and rowed as hard as they could.

The boats shot through the tunnel before it slammed shut, bringing the sailors up sharp with the west wind in their teeth. They had survived, but could now do nothing except let themselves be pushed back toward the shore and find firm ice to drag their boats up to and sit on until the wind and water invited them back.

The morning of July 11, they were still camped out on their ledge when a bear rose out of the water and ran toward them. Three men quickly got their muskets and prepared to shoot, all firing at thirty paces, all hitting, stopping the animal in its tracks. It dropped into the water, where it floated, senseless. Fat ran from the holes they had made in its body, blooming like oil on the water.

They had traded the cycle of shoveling snow and finding wood that had dominated their overwintering on Nova Zembla for an endless loop of shooting bears, dragging boats, and consigning themselves to the choking ice as they got weaker and weaker. Occasional meals of birds kept them going for the time being, but they wondered how much longer they could last. A piece of drift ice carried sailors over to the bear's corpse, and they dragged it back with them. Measuring its girth at eight feet, they smashed its teeth in. The gesture, transcending any actual need for protection, had become a ritual.

As the weather cleared, three sailors went to explore an island they'd spotted the day before. Once ashore, they saw more land to the west, and recognized it as Cross Island. Walking over fast ice between the two islands, the crew looked for any indication that Russians had been there that summer. No signs of human life were apparent, but the sailors found seventy burrow-duck eggs. Unable to carry so much fragile treasure in his hands, one sailor removed his pants and knotted the legs at the bottom. The men then loaded the eggs into the pants, which were hoisted gingerly between two of the sailors on the way back to the boats, while the third carried the musket, in case of bears.

Their roundabout walking route covered twenty-four miles in all. Twelve hours had passed before they returned egg-laden and partially unclothed to join their shipmates—who'd feared they were dead. Stranded for the time being on the ice, the sailors feasted, with enough eggs for every man to have several. The captain let them finish the last of the common wine, and they had three large servings each. They were still two thousand miles from home.

They spent the next four days hemmed in by ice, going ashore to get wood and look for interesting stones. On the third day, three men made their way to the island nearest them, shot a burrow-duck, and carried it back to share.

On July 16, a bear stalked them from the land side of their iceberg, so invisible against the snow that they didn't see her until the last minute. One sailor shot and hit her, but she ran away. The next day, the men were fretting about being delayed from setting out for so long and decided to go over to the close island to look for open water. Halfway there, they ran into the bear they'd shot the previous day.

As they approached, the creature made off, but one of the sailors followed and drove the pointed shaft of the boat hook he was carrying into the bear's body. The animal reared onto its hind legs, and the sailor tried to spear it again. Instead, the creature smashed the boat hook to pieces. The sailor fell down backward, and the bear turned toward him. The other two sailors came closer, and shot the animal in the torso, driving it down off its hind legs until it could barely move. They fired again then struck the teeth from its jaw.

The following day, three sailors went back to the island to hunt for a navigable route. It had been one year since they'd first caught sight of Nova Zembla on their voyage. At that time, they'd been looking northward in search of open water. Now they looked south, hoping to leave the ice entirely behind. From the highest vantage point, plenty of open sea was visible away from the coast. But the distance seemed so great, they worried it would be impossible to drag their boats that far.

When they returned and debated the idea with the whole company, the crew decided that they would try. They rowed toward the ice barrier, hosted the boats up onto the ice, emptied them, dragged the vessels almost half a mile to the other side of the ice, and then came back to carry their belongings across. Midway through, their strength nearly gave out. But they told themselves it might be the last time they'd need to move the boats this way, and they pressed on.

By the evening of July 18, they'd gotten the boats loaded and back into open water. They set out once more, but soon struck ice again. Only a day after they hoped they'd crossed the final ice blockage of their voyage, they once more found themselves hoisting their vessels out of the water. From atop their ledge, they could see Cross Island, now just four miles away. The next morning, seven men went ashore and climbed up to look out from its heights. To the west, they saw open water everywhere. The sailors hurried back to the boats with the good news, bringing a hundred eggs with them for good measure. Cooking their eggs quickly, they convinced themselves that maybe they could carry their boats one more time.

They dragged the vessels some five hundred feet and put them in the water. Suddenly, a gale rose up, quickly carrying them past and away from Cross Island. Forty-eight miles later, they passed Capo Negro, and by the evening of July 20, they'd gone another thirty-two miles and reached Admiralty Island. Hundreds of walruses lay out on an iceberg. The men took the boats in close and drove the animals from the ice into the water. Accustomed to their dominance over most Arctic life, the walruses swam toward the men, surrounding their boats. The creatures began making noises and seemed as if they might attack. In small boats, the bedraggled sailors could be easily overturned or sunk by a concerted effort from even one of the two-thousand-pound walruses. The crews and their boats were saved only by a strong wind, which let them flee from the fight they'd nearly provoked.

On July 22, they traveled a bracing sixty-eight miles without ice. They'd made such good time that the captain let his men go ashore in search of eggs. They came up empty-handed, but back at sea later in the day, they spotted a high cliff filled from top to bottom with birds nesting in crevices. The sailors killed twenty-two birds with rocks, and a nimble shipmate gathered fifteen eggs from nests before the captain urged the men back into the boats to take advantage of the steady breeze that still blew. After fair progress, they came to yet another cliff filled with birds and killed well over a hundred using stones and their bare hands.

Returning to the boats, the men in de Veer's craft found a strong, northwesterly wind had risen, and ice began crowding in. They tried to avoid the current as best they could, but they were pulled in with the smaller blocks and towering slabs of ice. Once caught in the flow, they saw open water closer to shore, and made their way toward it, finding the going easy again.

Thinking at first that de Veer's crew had gotten in trouble with the ice, van Heemskerck waited. But seeing that his mates had free sailing, he tacked to follow behind them. They eventually came to a good harbor where they could safely land and get wood for a fire to cook their catch.

They remained socked in without good sailing weather for three days, trying to take the sun's height with the astrolabe and searching for more eggs or valuable stones. On July 26, when the north wind held and the skies cleared, they finally set out again. But the going was hard. They had to sail sixteen miles offshore just to round a cape, borrowing the wind when they could and otherwise leaning into their oars. They cleared the point of the cape just after midnight and headed back toward land.

Taking in their sails, they rowed through massive shards of ice near shore all the next day, until they came to a broad stream flowing out from the land. They guessed (correctly) at this point that they were near Kostin Shar and wondered (incorrectly) if the water might flow all the way through to the sea on the eastern side of Nova Zembla.

They'd long since left behind terrain familiar to any of them, but they made good progress as they sailed. Van Heemskerck drove ahead of them at one point, before halting to rejoin forces and look for birds, which, unfortunately, were nowhere to be found.

July 28 brought fair weather again, and they continued to sail near shore, stopping not far from Mealhaven, where Barents's men had found buried sacks of grain on his first expedition north. They saw two ships near the point, and some men moving along the shore. They hadn't yet traveled the full length of Nova Zembla, but they'd rediscovered humanity.

Along with their delight at the presence of other people, they felt a parallel anxiety. It was obvious that between them, the two ships carried a minimum of scores of men—far more dangerous company than a dozen sick sailors in two ramshackle boats. And the castaways couldn't determine the nationality of their new neighbors, which meant they had no idea whether they'd be met with greetings or violence.

Consigned to their fate, they rowed hard against the wind and headed to land to find out. The men along the coast left their work and, unarmed, came to meet the Dutch boats. Van Heemskerck, de Veer, and all the sailors who could still walk climbed out to greet the strangers.

They were met with shock and pity. The sailors were Russians. Some had been near Vaigach Strait two years before, and recalled Barents's second expedition to the region. They'd seen the seven-ship convoy in all its glory and boarded one of the vessels. Those who'd been present then recognized van Heemskerck and de Veer.

Gone was the fleet. Gone were the proud Dutchmen on the cusp of sailing to China. Now they stood before the Russians in abject misery, consumed by scurvy. Approaching them with concern, the Russians delicately asked, *Korabl?*, the Russian word for ship. They had no Russian interpreter, but de Veer and van Heemskerck nonetheless knew the word. They tried to make signs to show that the ship had been trapped by the ice. The Russian reply was a phrase the Dutchmen recognized: *"Korabl propal?"* Yes, acknowledged van Heemskerck, the ship was lost.

Conversation was limited, but recalling the wine they'd drunk together before, the Russians asked what the Dutchmen were drinking now. One of the Dutchmen went to get water from their stores in the scute. The Russians tasted it and shook their heads, indicating that it was a sorry thing, indeed, to be reduced to drinking that. Hoping to learn of any other cures for scurvy, van Heemskerck moved closer to show them his mouth, with its loose teeth and diseased gums. They mistook the gesture for a display of hunger, and one of the Russians brought out food to share. He offered them an eight-pound loaf of rye bread and some birds. In return, the Dutchmen offered most of the last of the captain's tiny reserve of wine and half a dozen ship's biscuits—which some of the crew had grown too sick to eat.

The Russians invited them back to camp to sit by the fire, where the Dutchmen cooked biscuit porridge, looking for both sustenance and warmth. Across thirteen months, these Dutchmen had seen more bears than people, and even the population of their small human outpost had dwindled over time. Though they remained far from their own corner of the world, they were still astounded to have rejoined the company of the living.

On July 29, the Russians organized their provisions and equipment to leave. Before setting sail, they dug up barrels of whale oil

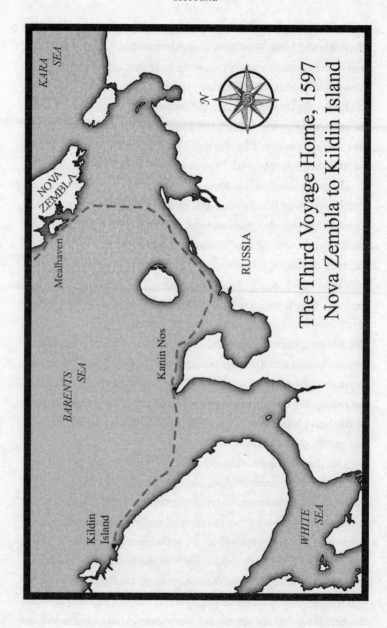

The Third Voyage Home, 1597
Nova Zembla to Kildin Island

KARA SEA

NOVA ZEMBLA

Mealhaven

BARENTS SEA

Kanin Nos

RUSSIA

WHITE SEA

Kildin Island

on the beach and loaded them into their ships. Seeing their new-found friends heading toward Vaigach Island but unclear of their final destination, the Dutchmen followed them. Bad visibility and the need to stay close to land meant that the scute and the rowboat soon fell behind. They turned to their planned route and watched the map, sailing between two islands until ice blocked their way once more. They turned around, working their way back to the islands for refuge.

The next day brought torrential rain and storms, delaying any possibility for departure. The sailors stretched their sails over the boats and huddled inside, but the makeshift tent failed to keep them dry. Unaccustomed to planning for rain, they had nothing else with which to protect themselves.

The last day of July brought clearer skies, however, and they took the opportunity to row from their island to one nearby where they'd seen two crosses. It had looked as if traders had visited, but any who'd come through were already gone. The men decided to go ashore and once there noticed something strange and startling in the pale landscape. Spoon-wort, a kind of grass, grew in vast quantities there. They hadn't seen a fruit or vegetable in more than a year, but quickly set to work eating the grass.

Spoon-wort, a low, creeping plant with round leaves, would later earn the nickname "scurvy grass." Common in parts of the Arctic for centuries, it contains large quantities of vitamin C. Nearly a century after Barents sailed, an Englishman would write a whole book in praise of scurvy grass—with descriptions of countless different internal and external uses for the plant. In time, it would become a popular treatment for scurvy internationally. The spoon-wort on Nova Zembla may even have been planted there by Russian traders for just this purpose. But however it made its way onto the island, van Heemskerck and his men saw it as a gift from God and believed that they'd been drawn there for the express purpose of finding it. On some level, they understood what they needed, pulling it out of the ground and eating it by the handful.

That morning, they'd been almost too weak to row their boats, but as they digested the grass, they began to feel better immediately. The sea, however, went wild again, threatening their boats so directly that they had to row them to the other side of the island for protection. When they got the scute and rowboat back to shore, they found more spoon-wort and continued eating. Some of the men who'd been unable to tolerate the ship's biscuit soon found they could eat it again.

But the remaining biscuit didn't last long. And though the worst symptoms of scurvy began to diminish, hunger took hold. They had a little bread, but it was moldy. A few of the men had the last of their cheese that they'd saved. There was nothing else left to eat.

On August 3, after three days with the spoon-wort, they decided to leave Nova Zembla behind and strike out for the Russian coast-line, for fear they'd starve where they sat. They set out in the morning with a northwest wind, but soon ran into their nemesis, as frozen walls surrounded them once more. After being certain that they'd freed themselves for the last time days before, they were devastated all over again.

Without a wind they couldn't sail, and so they had to make their excruciating way through the maze of ice by rowing. After hours of work, they slipped into the open sea. All the ice seemed to have vanished. They covered some eighty miles, and began to keep a watch for the mainland.

Instead they were met by ice, and a fierce cold returned. They'd imagined themselves leaving behind the most bitter elements of their journey—cold, impassable seas, and hunger—only to face them all at once after they'd come to think themselves safe. The scute could navigate, but the rowboat had more trouble, and couldn't find a way around the farthest point of ice. De Veer and his mates in the rowboat could see open water in the distance, but to get to it, they'd have to pass through the barrier of ice. Chief among their problems was finding a place to enter.

When they finally managed to slip into the ice belt, the route became a little clearer. They rowed themselves along the inside track in their misery of exertion, until they saw van Heemskerck round the ice from the far side of the current. As he came back toward them, de Veer's boat broke free from the ice, and they reunited, each halting success bringing them a little closer to home.

On August 4, they caught a good wind and rode it almost due south. As the sun crawled to its highest point for the day, they looked out and saw the coast of Russia. In their open boats, they'd sailed some one hundred twenty miles off the coast of Nova Zembla, and seven hundred miles from their cabin at Ice Harbor. The land was low and bare, and seemed as if it would be prone to flooding, but for the time being it was above water. They dragged their boats up onto the shore—the same connected earth that, if they could follow it far enough, would run right to their doorsteps in the Netherlands.

But the land they stood on still lay far from home. The men spotted a Russian boat, which they set sail to approach. As they moved alongside, sailors on the ship came above deck to talk to them. The Dutchmen cried out "Candinaes! Candinaes!" using their bastardized pronunciation to ask if they'd reached the cape they knew lay on the east side of the White Sea—a cape known to the Russians as Kanin Nos. Hearing the Russians' answer, they realized with dismay that their boats sat some two hundred miles farther east than they'd hoped.

Though they'd mostly sailed along the coast on their trip, which they could track with the map Barents had made, as soon as they left sight of land, the loss of William Barents as chief navigator had been felt. They couldn't find their location with any accuracy. They ended up spending the night where they were.

The next day, one of the sailors went to explore the area. He came back urging his shipmates to return with him. Not far inland, he explained, trees appeared, and the land turned green. There was game to shoot. The crew had been forced to fast for several days with only some moldy bread remaining, and each man getting only four ounces a day of that, roughly one slice. They wondered how soon they might die of hunger.

But on August 6, they decided to press on, to get past what they thought was the opening of a creek—but was actually the mouth of a river. They rowed for twelve miles against the wind until they couldn't fight it any longer. "Heartlesse and faint," they saw the land on the opposite shore still stretching northward out of sight.

The next day, the wind pushed them into the river, but they sailed eastward back the way they'd come until they emerged from the river mouth once more. They'd wasted days only to end up on the same spot where they'd first gone ashore. On August 8, the weather was no better, and the wind remained opposed to any forward motion by the boats. The vessels sat some distance apart, and in de Veer's boat, despair was mounting. Stuck in place without food all day, they waited for the wind.

August 9 was no better, and a desire for death began to take root. Two men went out from van Heemskerck's boat, heading for land. Two more went out from de Veer's boat, and together the four walked several miles inland. There they found a beacon and the corpse of a dead walrus. They dragged it back with them, thinking to use it to feed the men. But the corpse stunk ferociously, and their fellow sailors back in the boats thought it would surely kill everyone. In the end, they decided not to tempt fate.

The following day offered only dirty weather, pinning everyone in the boats. It was so miserable that the sailors kept their silence, knowing there was no good news to share by talking. On August 11, van Heemskerck announced they would set out again. Gerrit de Veer was brought into the scute, and a sailor from it was sent over to take his place in the rowboat. Though only in his twenties— at what should have been the height of his strength—de Veer had grown too weak to row, and the boat couldn't be managed with only the remaining men. Those who still could began rowing and kept at it until a hard wind rose up. The sails were struck, the oars came out, and once more they headed back to shore. But they'd managed to sail for most of the day.

Being ashore meant they might get fresh water, but on land, they couldn't find any. The weather was foul, and they laid the sails over each boat like a roof. Thunder and lightning plagued them, and they had no peace.

On August 12, the weather improved, and they spotted a Russian ship at sea headed toward them. Begging the permission of the captain, the crew took the boats into deeper water to meet with the visitors. Van Heemskerck went aboard to try to find out how much longer they had to sail to reach Kanin Nos, but the language barrier made it difficult to get any clear answer. Their hosts held up five fingers, and eventually the Dutch sailors understood there would be five crosses on it. The Russians demonstrated to them with a compass that the place lay northwest of their current location. Van Heemskerck made his way deeper into the ship and pointed at a barrel of fish, holding up a Spanish coin. They took the coin and gave him more than a hundred fish, along with some meal cakes. The Dutchmen divided everything evenly without regard to rank. Finally, they could eat.

A south wind helped them in their quest to sail west and north.

Thunder and rain overtook them, but the storm quickly died down and they sailed on. Later, they went ashore, with two men hunting for a good vantage from which to spot the point of Kanin Nos, where they would set out to cross the open sea. The scouts spotted a house with no one home and, looking along the coastline, felt sure that they'd arrived. The sailors took heart in the news, and went back to their boats again and began to row along the coast. They spotted the wreck of a Russian ship and another house, which they stopped to investigate. Again, they found no one home but discovered some spoon-wort along the way, which they brought back to share.

They had a strong wind from the east, which worked in their favor. In the afternoon, they passed the outcropping of land they'd seen and were filled with anticipation. They met up to plan, and boarded each other's boats. Carefully dividing all the candles, supplies, and cargo, each man also gathered his own possessions to keep with him, in case the boats lost track of one another other while crossing the White Sea. They hoisted sails and set out across open water, braced for the battering they knew the boats would take.

Near midnight, a violent storm struck. The scute took in some sails to better weather it, but the rowboat did not. In the squall, the two boats were separated. They continued to sail into the morning on August 14, and as the weather cleared, the men in the scute spotted the rowboat in the distance, but they couldn't catch up. Fog set in, and they gave up any hope of pursuit. But they thought they'd keep to the planned course and try to meet up with the other boat on the far shores of the White Sea.

De Veer's company sailed west northwest, as well as they could reckon it, with their sails up for the first part of the day. But when the wind turned against them, they returned to the oars, though they wondered at their compass and whether it might be malfunctioning. The next day, they continued alternating oars and sails and caught sight of land. As they worked their way toward the coast, they noticed six Russian ships.

They sailed up to ask how far it was to Kildin Island. Though the language barrier that had plagued them before was just as much in effect, the Russians seemed to say that Kildin was still some distance away. Not only that, but they seemed to be claiming that the Dutchmen were still on the eastern side of the river mouth at Kanin Nos—that they hadn't crossed the White Sea at all. Their hosts spread their hands out to show great distance, adding that the foreigners' boat was too small to survive the trip.

The baffled sailors asked for some bread and were given a dry loaf that they ate with pleasure. Parting company with the Russians, they concluded that there must be some confusion. The men refused to believe that they hadn't reached the White Sea yet. On August 16, catching sight of another ship that appeared to have sailed out of the open water they'd just left behind, they approached it, rowing with great difficulty.

Once aboard the ship, the Dutchmen asked how far they were from Kildin Island. These Russians had the same response as the others, indicating that they were still on the east side of the river mouth. Again, the Dutchmen refused to believe it.

Van Heemskerck asked the Russians for some food, and after paying for it, headed back to the scute. Preparing to leave, they readied themselves to work against the current to get clear of the inlet they'd entered to catch the ship. Knowing that high tide would soon ebb, the Russians sent two men in a small boat, inviting the foreigners to come back to the ship. They passed along a loaf of bread, which the castaways thanked and paid them for, but van Heemskerck wanted to set out without delay. Seeing that they remained unconvinced, the Russians aboard the ship dangled bacon and butter to beckon them.

After their guests climbed aboard again, the Russians showed them their location on a map, insisting that they were still east of the White Sea. Pulling out their own chart, they examined it with the Russians, who infected their guests with doubt and alarm. They seemed to have hardly covered any distance at all.

Realizing that they'd yet to find the White Sea, the Dutchmen dreaded the idea of crossing such a large body of water in unsafe boats with no food. Remembering how much difficulty their friends in the missing rowboat faced navigating choppy water compared to those in the scute, they feared for their mates' lives.

Van Heemskerck bought three sacks of meal, two and a half sides of unsliced bacon, a pot of butter, and a small cask of honey to share between the scute and the rowboat, whenever they might find it again. They set out with the ebb of water in search of their friends and Kanin Nos, which would appear if ever they reached the eastern shore of the White Sea. They thought at one point that evening that they might have arrived, but when they got to the place where the land had seemed to stretch out into the water, they found it merely fell away to the northwest. Making little progress against the tide, they stopped and cooked a pot of meal with bacon fat and honey, practically a Twelfth Night feast. But there was no sign of the rowboat or their shipmates—who were surely just as hungry but rowing with no meal at hand—which dampened some of their delight.

Early in the morning of August 17, they met up with another Russian ship. A sailor aboard brought them a loaf of bread. Through gestures, the Russians seemed to say that the rowboat with their friends had been spotted. Hardly believing it, the Dutchmen tried to get more information, and learned that seven men had been aboard. Just the day before, the Russians had sold men in the rowboat bread, meat, fish, and other food. Recognizing a compass the sailors had likely traded for provisions, van Heemskerck and the crew of the scute were overjoyed that their mates were on the same course and also had at least some food.

They left quickly, in the hope of catching up to their friends. They rowed hard, fearing that the other crew might not have enough to eat to keep going for long. Near midnight, they saw current flowing ashore, and stopped to collect fresh water on land, gathering more spoon-wort leaves as they went. The next day, they rowed along the coastline after hoisting their anchor—which in their poor scute wasn't a real anchor, but a stone at the end of a rope. At midday, they looked out and saw an outcropping of land stretching into the sea, with what looked like, perhaps, several crosses. As they drew near, the crosses grew clearer. Five crosses. They'd finally reached Kanin Nos.

While making final preparations to sail one hundred sixty miles across the sea without any possibility of landfall, they noticed one of their casks of water had leaked and sat nearly empty. They looked for a place ashore to refill it, but the waves were too brutal, and they gave up all thought of stopping. Under a favorable northeast wind, they set out and passed Kanin Nos with the evening sun. Their sails carried the boat into the night and all through the next day, with only an hour and a half of rowing. They sailed throughout the evening, and as the sun rose the next morning, they heard the sound of waves breaking on the shore. After only thirty hours out of sight of land, they caught sight of cliffs, hills, and mountains that seemed a world apart from the low terrain they'd seen on the eastern side of the sea. The passage on their homeward journey that they most feared had turned out to be the easiest.

But the fair wind abandoned them there, and they decided to try to get to shore. The scute now sat along a more heavily traveled sea route, and sailors found warnings and guides posted along the coast to help direct them. Heading to a sheltered stretch near shore that was posted as safe, they saw a large Russian ship at anchor and a string of houses by the shore. As they rowed quickly to it and dropped their anchor stone, it began to pour. Stopping to cover their boat with a sail, they went onto the beach and up to the houses. The sailors there invited them into their quarters. The visitors were made welcome: their soaked clothing was dried, and their hosts cooked fish to feed them.

The cabins housed thirteen fishermen, who went out each day to fish under the direction of two supervisors. They had nothing of their own, and only lived off fish and more fish. With little else to offer, they invited van Heemskerck and de Veer to spend the night in their cabins. Van Heemskerck declined, and said he would stay with his men, but de Veer, who had been very ill, slept away from his mates that night. Along with the Russians were two Sami men, three women, and a child, who seemed to be fed only on the scraps provided by the Russians.

Night turned to day on August 21, and it continued to rain. Van Heemskerck bought fresh fish for the crew. Cooking some of their ground meal with water, the men ate until they found their stomachs full. They felt at peace. As the rain diminished, the boat stayed at anchor, while some of the crew went farther inland for more spoon-wort. Up in the hills, they spotted two figures coming toward them, and wondered if the region was somehow more populated than they'd realized.

Making their way back toward the scute, they realized the two men they'd seen were following them. As the pair came in sight of the scute near the shore, the men grew excited, and de Veer's company realized it was their mates from the rowboat. They'd landed somewhere nearby and met by chance. The new arrivals were hungry but had no money with which to buy fish. So they'd planned to trade a pair of pants for food. The sailors from the scute fed the mates who hadn't eaten, and everyone had as much water as he wished.

The rowboat joined the scute the next day, and the Dutchmen asked the Russians to bake a sack of meal into bread for them. As the fishermen brought their catch back, their guests bought four cod from them and cooked it. The Russians gave them extra bread for good measure.

On August 23, the cook kneaded and baked bread from another sack of meal, and they prepared to set sail again. Van Heemskerck gave the Russians a good tip for their generosity, and also paid the cook for his help. The Russians asked for some of the crew's gunpowder, which they were given, the Dutchmen having little use for it away from the bears. Before leaving shore, they divided the remaining meal between the scute and the rowboat, so that neither crew would go hungry if the sea separated them again. They headed out near evening and kept close to land.

The next day, they got as far as a group of seven islands marked on their map. Asking the fishermen they met to point the way to Kildin Island, the fishermen directed them westward, throwing a cod in their boat as a gift, and they sailed on. They had no plans to search for passage home at Kildin, but Kildin was on the way to Wardhuys, where they felt certain they could find a ship to carry them.

Later the same day, farther along the coast, a boat of fishermen rowed out to ask where their ship was. Having some experience by then, the Dutchmen were able to say *Korabl propal*, and it was understood that their boat was lost. The fishermen seemed to say

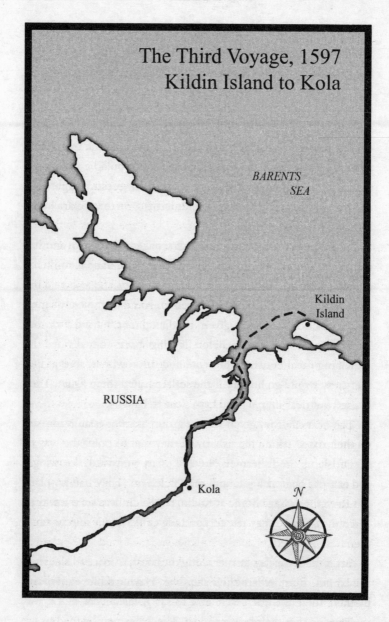

The Third Voyage, 1597
Kildin Island to Kola

*BARENTS
SEA*

Kildin
Island

RUSSIA

Kola

N

that there were Dutch ships not far away at Kola, a trading town at the mouth of a narrow inlet on the mainland. But van Heemskerck put no stock in the idea and kept his plan to sail to Wardhuys.

They sighted Kildin on August 25, and navigated between it and the mainland, arriving at its western end. They anchored near a Russian ship they spotted there, and van Heemskerck went ashore to nearby houses. The Sami who lived there told him that three ships from the Netherlands were at Kola, and two would be setting sail that day. Van Heemskerck returned ready to leave, thinking that the crew might be able make a run to Wardhuys to intercept the vessels. But as they tried to go to sea, the wind beat them back with such force it was clear it would be foolish to try to sail that night. The sea went hollow again, and they feared that each wave might swamp the boat. They took refuge by sailing behind some cliffs toward land, and there found a house with three men and a large dog. Sharing some details of their misfortune, they heard for the third time that Dutch ships were slated to sail out from Kola that day.

Asked if someone could take the Dutchmen to Kola by land, the three men declined. But they offered instead to escort their guests over the hill to a Sami man who might agree to take them. He proved willing, and van Heemskerck paid two Spanish dollars to send one of the Dutch sailors to Kola.

On August 26, the skies had calmed. They pulled the boats ashore and emptied them, to air out their belongings. They carried food to the Russians' house and cooked a meal. From here forward, they realized, enough coastal dwellers and fishermen would appear along their route that they no longer had to hoard their food. They'd begun to eat two meals a day, and were drinking Russian kvas, an alcoholic drink made from fermented bread. They collected blueberries and blackberries, further healing the damage scurvy had done.

They still slept in the scute and the rowboat at night. On August

27, nasty weather out of the north compelled them to haul their makeshift homes farther up the shore to keep the high water from dragging their vessels into the water and flinging them back onto the rocks. After moving their boats, some men then went up to the Russians' place to sit by the fire. While they were gone, the water roared up and pulled both boats into the sea.

Two men had remained in the scute and three in the rowboat, but they couldn't control the boats. All they could do was to try to keep them from being smashed to pieces. Eventually, the storm subsided and the boats were rescued from the surf, but in the meantime, the entire crew was out in the deluge for hours while it rained across the evening and all through the night into the next day.

When the weather subsided somewhat on August 28, the boats could finally be pulled ashore again, and emptied once more. The rain persisted, and the crew laid the sails over the gunwales, to make a shelter for themselves. They wondered what had become of the man sent to Kola with his Sami escort, and whether any Dutch boats had been found there. They ate berries, and, looking for some sign, watched the hill their mate had gone over when departing. But the day passed without news.

Night arrived, and morning came behind it, bringing clear weather. The sailors cooked their meat at the Russians' fire and ate, then headed back to their boats to settle in for the night. On their way to the shore, they spotted the Sami man they'd sent out coming down the hill alone. They wondered what had happened and feared for their mate. Asked for an account of the trip, the man handed them a letter addressed to van Heemskerck.

The letter was written by someone who obviously knew them well. The author was amazed by their arrival at Kildin and Kola, because all of van Heemskerck's and Barents's men had been assumed dead. He was "exceeding glad" at their arrival, and would soon appear with food and anything else they might need, and would take care of them. The missive was signed "Jan Cornelis Rijp."

They were sure it couldn't be their Rijp—the Rijp whose ship had set out with them from Amsterdam in spring of the prior year. Rijp had gone his separate way more than a year ago at Spitsbergen, some five hundred miles away. Who had any idea where he was now? For all they knew, he might have sailed north and arrived in China, and still be there. He might have turned for home. Most likely, he was dead. Perhaps it was some other man also named Rijp.

But the letter was a wonder, and it seemed nothing bad had befallen the shipmate they'd sent off with the Sami escort. They paid the man his fee, and gave him hose, britches, and other clothing, so he could dress like a Dutchman any time he liked.

They sat debating the identity of the author of the letter, and van Heemskerck recalled that he had among his papers something that had been written by Rijp. Going to fetch it to compare the handwriting, he put the two papers side by side. The sailors then realized it was one and the same, that this Rijp was their Rijp, the one whom Barents had quarreled with, and to whom they'd said their goodbyes before all the misfortunes and salvations of the last year had taken place.

Yet some men refused to believe it was true, right until the moment that a Russian boat appeared the next day rowing to shore with Jan Cornelis himself aboard. They greeted each other as only those who have mutually risen from the dead can. Rijp had brought with him a barrel of Swedish beer, wine, and liquor, as well as bread, meat, salmon, sugar, and other food. They gave thanks to God in his mercy for preserving them all.

They learned that after William Barents and Rijp had split the fleet at Bear Island more than a year before, Rijp had tried again to sail due north toward the pole. After again being foiled by ice, he had safely returned to the Netherlands that summer. The following spring, he set out on a trading mission to Kola. He hadn't been searching for Barents, van Heemskerck, and the other men, who were assumed to have perished. They'd found each other by sheer chance.

On August 31, a wind rose up which could carry them from Kildin Island to the town of Kola. Rijp's ship was busy taking on cargo and stores for the return voyage and couldn't come to them. They'd have to make their way into the long mouth of the bay down to Kola. They thanked the Russians for their food and drink and hospitality, giving them money as they said goodbye.

They set out at night and high tide, arriving the next morning on the west side of the river mouth that would lead them to Kola. They cast out their stones for anchors and waited there during low tide. When the water returned, they set out once more. They repeated the process until the morning of September 2 brought them in sight of trees and buildings, where they stopped for a short while in what felt like a return to human society. Continuing another twelve miles, they came to the ship captained by Rijp and stopped to visit and have a drink with some of the men who'd been in the fleet that sailed with them to Amsterdam the year before.

After a time, they rowed on to Kola itself by evening, where some kept an eye on their cargo and others went ashore. They came back with treasures like milk for the men who'd stayed in the boats. They were by no means in Amsterdam—the culture remained very foreign to them—but for the first time they felt safe.

They unloaded the boats the following day, and spent the next week eating and sleeping, making some semblance of recovery. On September 11, with the blessing and permission of the local representative of the grand duke of Moscow, they hauled their rowboat and scute into the merchants' house, where they'd be left as a memorial to the voyage, having sailed sixteen hundred miles from lands the sight of which had never before been recorded.

On September 15, all the men but van Heemskerck climbed aboard a Russian ship, which carried them and their belongings, along with the surviving cargo, to Jan Cornelis Rijp's vessel, which was anchored outside town. At noon, the ship weighed anchor and sailed until they'd passed the narrowest section of the river, where they waited for Rijp and van Heemskerck to come from town to join them.

Rijp's ship floated out of the mouth of the Kola River around six in the morning on September 18. Two days later, they landed at Wardhuys, where William Barents had told van Heemskerck he might one day return to walk on shore. Those who'd sailed with Barents had three more weeks to recover from scurvy and build their strength for the voyage home, while Rijp did business and made arrangements to take on more cargo. They left on October 6, with a passage over Norway so familiar and unextraordinary that it hardly merited more than the captain's record of the wind and weather. Across three weeks, they sailed back over the farthest northern points of Scandinavia down into the North Sea, finally reaching the Dutch coast on October 29 at the mouth of the Meuse River. They continued on to Delft, making their way north to the Hague and Harlem before finally reaching Amsterdam.

As they sailed back into the Zuiderzee, following a tightening spiral until they came to the fortifications of the harbor, it was as if they were undoing the burden of the misery and sorrows they'd collected on the way out. They'd finally come home. But the weight of the loss at sea of the carpenter in the first weeks on Nova Zembla, of a second mate in January, of Claes Andries and his nephew John—and especially of William Barents—couldn't be undone. Not everything would be restored.

Near noon on November 1, the twelve surviving sailors from William Barents's ship arrived in the port of Amsterdam. As they caught sight of the city, they stood on the deck of the ship in the same clothes they'd worn since leaving the summer before, the same leather shoes that had frozen to uselessness in the Arctic winter and then thawed again in spring. On their heads sat the white fox-fur hats they'd stitched together in the cabin on Nova Zembla. Aside from their hats, they reentered the world more empty-handed than they'd left it, their survival story the only thing they had to share other than the pelts of dead animals.

Once ashore, they headed to the house of Peter Hasselaer, the merchant who'd fitted out their ships the year before for the city of Amsterdam. Believing Barents and van Heemskerck and all their men to "have been dead and rotten," those who greeted the returning sailors were amazed by their reappearance. The news spread quickly through town. The sheriff and two town council members came to get the sailors and escorted them to the Court of the Admiralty.

In front of the visiting lord chancellor of Denmark—and the leading men of Amsterdam who'd gathered for dinner—they told their tale. They and no one before them had sailed north to Spitsbergen, split with Jan Cornelis Rijp, and turned east toward Nova Zembla. Their ship seized by ice, they sat stranded all winter in great hardship and danger with no hope of rescue. They alone had to fight polar bears and poisoning and snow and wind for nearly ten months ashore, and even after surviving polar night, had to find their way back in open boats after suffering so much and almost losing their lives. The way home had been even more bitter than the voyage out and their overwintering ordeal, yet they'd not only survived but returned to tell the story.

In the centuries that followed, master of the heroic couplet

Alexander Pope would mention Zembla in his eighteenth-century poem "An Essay on Man." Pope notes that vice, like the idea of the distant north, is something all humans are acquainted with while still believing that it's located at some distance from themselves. Everything is relative, depending on how far gone you are—or how far you go. From Scotland, north means the Orkney Islands, but on the Orkneys, north means faraway Zembla.

In the "Battle of the Books" from 1704, Jonathan Swift set a "malignant deity called Criticism," who was both daughter and wife to Ignorance, as far as possible from any human civilization on a throne in a cave on the remotest heights of Nova Zembla. Charlotte Brontë, in the opening pages of *Jane Eyre* in 1847, mentioned both Spitsbergen and Nova Zembla to summon the desolate feeling of the far north. For *Twenty Thousand Leagues Under the Sea* in 1870, Jules Verne would trap his narrator near the end of the novel in a runaway submarine veering off toward Spitsbergen or Nova Zembla. "I could no longer judge of the time that was passing," writes the narrator. "The clocks had been stopped on board. It seemed, as in polar countries, that night and day no longer followed their regular course."

In 1962, Vladimir Nabokov would publish a novel with a mad narrator named Charles Kinbote, who believes he is the exiled king of Zembla. Salman Rushdie included a perpetual night covering part of the world in his 1990 novel *Haroun and the Sea of Stories*, with a phrase in his dedication mentioning

> *Zembla, Zenda, Xanadu:*
> *All our dream-worlds may come true.*
> *Fairy lands are fearsome too.*

Zembla would never lose its hold on the literary imagination. Writer William Boyd would fold the idea of Zembla into his 1998 novel *Armadillo*, coining a new word from it. Serendipity, he noted, came from Horace Walpole, who invented the word out of a folktale about the island of Serendip (now Sri Lanka), in which "heroes were always making discoveries of things they were not in quest of . . . serendipity, the faculty of making happy and unexpected discoveries by design."

> *So what is the opposite of Serendip, a southern land of spice and warmth, lush greenery and hummingbirds, sea-washed, sun-basted? Think of another world in the far north, barren, icebound, cold, a world of flint and stone. Call it Zembla. Ergo: zemblanity, the opposite of serendipity, the faculty of making unhappy, unlucky, and expected discoveries by design.*

Along with making Zembla legendary, Barents and his men would themselves become famous. By 1600, less than four years after their frozen Twelfth Night feast on Nova Zembla, William Shakespeare would write his own play about the same holiday. *Twelfth Night* likewise tells the story of a world turned upside down on this strangest of holidays, in which the high are brought low and everything spins topsy-turvy. A not-quite-dead dead twin, cross-dressing, and a plot nested around switched identities lead to a comedy of errors with its own holiday feast at the center—and a reference to Barents. When one character earns another's disdain, he's told, "[Y]ou are now sailed into the north of my lady's opinion; where you will hang like an icicle on a Dutchman's beard." In the space of a handful of years, the tale of Dutchmen covered in ice at the northern edge of the world would cross borders to become an international cultural touchstone.

Gerrit de Veer's account, *The Three Voyages of William Barents to the Arctic Regions,* would enter the historical record as a survival story beyond compare. His narrative would reinforce the mystery that the far north had held since before even the days of Pytheas sailing from Marseille to the Arctic Circle nearly two thousand years before. The hand-to-hand combat with bears, the trapping of foxes, the hewing of shelter from unyielding ice—and beyond everything, the misery of the men—made the tale irresistible. Within a year, it would arrive in Dutch, German, Latin, and French: an Italian edition would appear months later. An English translation would find publication in 1609.

In time, through their winter ordeal, their temporary island home would come to symbolize the frigid North and a place of untold suffering. Because of Barents and his men, Nova Zembla would represent the impassable, opaque, and unconquerable Arctic.

Barents's expeditions bound triumph and tragedy inextricably together. Though the sailors hadn't signed up or sailed in the name of science, their discoveries would change the understanding of barnacle geese, of mirages, and the very geography of their planet, with observations that would take centuries to be fully understood. Yet the crew's final voyage would also mark the permanent opening of the far north to Europeans, launching a harvest and devastation that would never stop.

Triumph and tragedy were also writ small on the lives of the sailors themselves. Despite all the luck, skill, and heroics by which they managed to save one another again and again, the men who survived a winter on Nova Zembla could neither rescue, nor preserve, nor even bring home the body of their peerless navigator. But because of their survival, William Barents would become immortal.

CODA

The Shores
of Nova Zembla

Standing on the deck of a boat passing by Russia's Kildin Island, I felt the four centuries since William Barents had sailed across this sea to Nova Zembla vanish. As our small company of ten set out on a fifty-nine-foot boat in August 2019 to follow in Barents's wake, the captain, Mikhail, suggested we fill up on food while we could. Once we left land behind, he told us, we probably wouldn't feel like eating.

I'd never felt sick on any boat before, but Misha was right. Less than a day's sail from Murmansk, Russia's northernmost city, the crossing became a nausea-inducing pilgrimage—even for those not prone to seasickness. Waves seemed to tilt the boat in several directions at once, and the churning of the sea couldn't be expected to fully stop until arrival at the far shore. Crossing the Barents Sea, I spent nearly four days miserable in every position unless I stood on deck looking at the reassuring stability of the horizon.

For almost a decade I'd been hoping to go to Nova Zembla to see the ruins of the cabin where William Barents and his fellow travelers overwintered in 1596. But by day four, in the throes of seasickness and wondering if it would continue for the whole expedition, I found myself thinking in the middle of a sleepless night that it would be fine if a storm came and the crew decided to turn around and head back to the mainland.

Luckily, my nausea vanished as soon as we came in sight of land. And as if to redeem the queasy passage, it soon became apparent we were echoing Barents's Arctic travels in more ways than one. Not only were we sailing where he'd sailed, in a boat of almost exactly the same length (if not the same volume), but we were about to relive parts of his voyages.

Barents reached the small outposts of rock and moss that he'd christened the Orange Islands in 1594 on his first voyage. Gerrit de Veer wrote of the explorers cresting Nova Zembla's northern coasts and seeing "about 200 walrushen or sea horses . . . a wonderful strong monster of the sea." When our 2019 expedition reached the same island, we spotted countless walruses lolling ashore at the same spot. Sasha, one of the crew members, pulled out an accordion-like instrument known as a garmonica and began playing a haunting Soviet-era waltz. Dozens of walruses swam out to meet us, watching the performance with fascination, and snorting at us in response.

Even the heavens and air re-created the past for us. Looking back at the islands after we sailed by, the crew watched the square-cornered plateau of what we called "Big Orange" Island suddenly change shape. As more of us came on deck and grabbed cameras or binoculars, the flat surface of the island sprouted a skyline of buildings. Something like smoke rose above one of the tiny stone outcroppings nearby. Someone eventually recognized the puzzling vista as a mirage. And I recalled the Nova Zembla effect, another more profound kind of optical displacement that left Barents and his men confused and unsettled during their overwintering on the islands.

Half the castaways' voyage home seemed to have involved pantomimed conversations with Russian sailors. Though my friend Tatiana had come along as an interpreter, I was caught alone many times and had to communicate with even less functional Russian at my disposal than Barents's crew had. Just as had happened with van Heemskerck and his men, my Russian sailors tried to inform me of any number of things that I tried and failed to understand along the way.

At the same cliffs where de Veer described sailors stealing eggs from nesting birds, we, too, were able to step quietly and slowly up within arm's reach of a variety of island residents, from cartoonish puffins to sleek murres. And when we landed at the site where Barents's cabin had stood more than four hundred years ago, long timbers from the shelter lay in a rectangle on the original site, though they'd been moved and put back repeatedly since the cabin's rediscovery in 1871. Hundreds of artifacts have been collected from the earth there. I visited some relics in Amsterdam, on Spitsbergen at Svalbard, and in St. Petersburg, Russia—and no doubt walked over and around many that had woven their way into the soil of Nova Zembla itself.

On the way home, we repeated a part of Barents's trip that we hadn't intended to duplicate. Hours after we left the Nova Zemblan coast to set out across the Barents Sea, the boat's engine broke down. Suddenly, like Barents, we had little idea how long it would take to get to our destination, and had to sail nearly perpendicular to the wind when it ran against us to make the smallest forward progress. We continued using only sails for over a week, dependent on moving air to push us forward, and more than once sitting motionless in its absence. One forlorn day, we watched birds next to us in the water paddling with their feet, moving faster than the boat.

Despite so many of the elements recorded during Barents's voyages persisting into the twenty-first century, Nova Zembla today isn't entirely the same. The sticks, branches, and whole trees that littered the shore in Barents's day are now just as likely to be real litter: trash piled along the shoreline. The plastic flotsam that has washed up has changed the landscape, though perhaps not as much as the glaciers and ice that have retreated or vanished—a gradual reduction visible in satellite imagery over the last several decades, with a dramatic drop-off after 2006.

Our response to the Arctic was different, too. We didn't need the birds or their eggs for food. We put on a concert for the walruses, instead of trying to kill them or steal their tusks. Though we saw just one polar bear, as opposed to the dozens encountered by Barents and his men, if we had somehow managed to kill it, even in self-defense, it would've provoked a serious investigation. We had no need for shelter off the boat, and could try to minimize our presence. We even tried to avoid crushing the sparse, low-lying vegetation, which can take years to recover.

Our journeys differed, too, in that once we'd crossed the Barents Sea coming home, we bypassed Kildin Island and headed straight to the port of Murmansk, where our company disbanded, and we went our separate ways. But van Heemskerck and the surviving castaways sailed together all the way back to the Netherlands, telling their story as a group in more dramatic fashion to the notables of Amsterdam, who'd gathered for a feast.

After that dinner on November 1, 1597, the men who lived in the city returned to friends or family. Those who were from other towns found local lodging until they could be paid and go home. Though their work had been the engine of the voyage, providing the ingenuity and muscle for everything from building the cabin to rowing and dragging boats to trapping foxes, most of the sailors who'd joined Barents on his third journey north, once home, disappeared from the public record.

Jacob van Heemskerck, however, would remain in history's sights. Six months after his return from the Arctic, he sailed with the Verre Company to the East Indies. Well into the voyage, van Heemskerck became commander of the fleet. The prior expedition to the East Indies, which had set out in the spring of 1595, hadn't taken aboard any fruits or vegetables to prevent scurvy. But on this voyage, the Dutch were more prepared. The ship they sent out on the same route in 1598 would carry lemon juice—and lose only fifteen men.

Van Heemskerck later sailed to the region as commander of the fleet and helped shepherd the new Dutch nation as it supernovaed into a vast empire. In less than a century, the goods shipped by Dutch traders would eclipse the combined total of Spain, France, England, and Portugal, with several other European powers thrown in for good measure.[1]

Just as he'd outlasted his time in the Arctic, van Heemskerck would survive his southern voyages and return home to take part in the war against Spain that would continue, at greater or lesser intensity, for another four decades. As admiral, he'd lead the Dutch navies against the Spanish fleet near Gibraltar in 1607, dying in battle after losing a leg to a cannonball.

Like Barents, van Heemskerck became a martyr for his country. But in the end it was Barents's name that surpassed even van Heemskerck's. Though Barents never gained fame in battle and never found a trade route to China, he had planted a seed for a new kind of explorer, one whose fame lay in a combination of knowledge and endurance rather than martial glory. It still was a face of nationalism—merely the softer side of the imperial project—but it was a deeply human face.

William Barents would become less and less real over time. The gaps left by his biography, and his death, create an emptiness that makes it possible to project or reflect whatever the viewer wants to see.

Yet every famous Arctic explorer who endured horrifying ordeals, every adventurer to the North whose story became a bestselling book, every voyager vowing to fill in the map for national glory, every polar adventurer whose exploits were recorded with the newest technologies—from books to telegrams to photos to radio broadcasts to phones to satellite links—has walked in the path first blazed by William Barents.

In later centuries, the failure to establish habitable colonies or make successful trade missions wouldn't count against intrepid explorers. From a monetary perspective in Barents's era, however, his final voyage was a disaster, so much so that when his wife applied for a widow's pension from the council of Holland, asking for support for herself and the five children her husband had left behind, she was refused.[2]

The Dutch didn't immediately give up on the Arctic route to China, but a northeastern passage was centuries away, and the

Dutch wouldn't be the first to find it. It might be tempting to imagine that by *not* breaking through to China, Barents and his men greatly delayed the opening of the Arctic, but soon after Barents's death the high Arctic became a commercial focus for a very different reason. Within fifteen years, Western European nations would establish whaling on the coast of Spitsbergen, and the search for a northern route would get sidetracked as huge profits were wrung from whaling, nearly wiping out the North Atlantic right whale.[3] In time, European hunters and trappers would also join the Russian, Nenets, and Sami seasonal visitors to Vaigach Island and Nova Zembla.

The high Arctic hadn't been conquered, but it had been infiltrated and would never close. The process begun by the same States General that sent Barents north—the kind of process that had already led to the exploitation of the Americas over the prior century—would in time become part of the division and conquest of every region of the globe. Barents's overwintering coincided with the end of any world that might have escaped the boot heel of colonialism.

William Barents had sailed north with an idea that he would find a warm polar sea. But standing on the shore looking out from near the ruins of Barents's cabin on a summer day, I put my fingers in the water to find that it's achingly cold. When the wind stirs, this spit of land sticking out into the sea quickly becomes a numbing, desolate place. How much grimmer the view must have been during Barents's months there, when great fragments of ice pressed in all around him.

The land is now a Russian Arctic preserve and still very isolated, though each year there is talk of restarting a defunct cruise to bring tourists to the ruins of Barents's cabin and abandoned research stations.

Dutch experts in shipbuilding and navigation came together in recent years to build a replica of Barents's ship. Using relics of typical yachts of the day in combination with the original illustrations published with the narrative of the voyages, shipwreck expert Gerald de Weerdt and mechanical engineer Koos Westra have guided a group of volunteers in the hand-built construction of a vessel that duplicates the details posterity has preserved about the original. The builders say they may try to sail it from the coast of the Netherlands all the way to the cabin at Ice Harbor, just as Barents did. If they do, they hope to avoid the ice that plagued Barents. But even with the challenges of sailing a historical replica, the trip will be easier than it once was.

The fate that awaited Barents on Nova Zembla was part of a story that had been unfolding for more than a century before he set out and would continue for many more. The idea that Barents had started with—that navigation to China was possible, that by turning his ship farther north and avoiding the fate of prior voyages, he might crest the world and find an open polar sea—doomed his voyage.

Yet, strangely enough, he was perfectly correct in his assumption. The world to which he belonged set machinery in motion that can now be slowed but not reversed. With some consistency, snow and ice surveys project that by 2040—perhaps as early as 2030—there will be no ice left at the North Pole in summer. By August 2017, the planet had changed so much that a Russian gas tanker equipped for Arctic voyages could travel for the first time without an icebreaker escort, sailing a northern route from Norway to South Korea in two-thirds the time required for the traditional route through the Suez Canal. The open polar sea Barents had forecast will soon exist every year during the hottest months. And the planet will continue to warm.

This stupendous change will be the end result of a process in which Barents and his Arctic expeditions were in some ways the opening salvo. Though they returned with a dramatic tale of uninhabited lands and scientific insights, their ships still rode the wave of a tide that would unleash destruction as powerful and enduring as any force in human history.

The sea free of polar ice that the Greeks had deliberated over and Barents's own mentor had insisted was real wasn't just a figment of their imaginations. The open polar sea that Barents had imagined, the idea for which he'd risked everything, has finally come to pass. He just sailed four hundred years too soon.

Acknowledgments

A book goes out under its author's name, but a work of narrative historical nonfiction represents the effort and aid of countless people. To name some is surely to forget others who also deserve mention. But here are some of the key people with a role in helping to bring *Icebound* to life. I'm grateful to each one in different ways.

Rick Horgan, who acquired *Icebound* for Scribner and was my partner in crime, has an eagle eye for noticing what's missing in a manuscript. I'm beyond grateful to Scribner publisher Nan Graham for seeing the potential in the book proposal, as well as Beckett Rueda, Laura Wise, copy editor Jane Herman, and the design, publicity, and marketing teams, for their help in delivering this book into the world.

This is my third project with Katherine Boyle at Veritas Literary as my agent. Without her, none of this would have happened.

The involvement of my dear friend Beth Macy during the writing process was, as it has been for the last decade, a gift. Vanessa Mobley, Michael Robinson, Blair Braverman, Anna Badkhen, and Dan Vergano were all early readers of my book proposal or the full manuscript. Each one gave invaluable advice on improvements. Graphic artist Robert Lunsford was generous with suggestions on improving the maps in this book.

Acknowledgments

Translators were central to this project. Tjitske Kummer translated the Arctic voyages of Jan Huygen van Linschoten from Dutch, in what may have been the first such translation into English in history. She was also critical in helping to compare the English and Dutch editions of Gerrit de Veer's diary—which appears not to have been retranslated and republished since its initial appearance in English more than four hundred years ago (which explains the archaic spellings in the excerpts quoted here). Robert Neugarten translated scholarly papers on Barents and van Linschoten for me on very short notice.

Several scholars and researchers were unbelievably generous with their time—meeting me on days off, taking trains to accompany me to storage depots in other cities, copying materials, and answering questions ad infinitum. I'm particularly indebted to Diederick Wildeman at the Netherlands' National Maritime Museum, Jan de Hond at the Rijksmuseum, and Russian Arctic explorer Pyotr V. Boyarsky, who met with me on short notice in Moscow for several hours to talk about his work at Barents's landing sites. Historian Anne Goldgar, a longtime investigator into all things Nova Zemblan, caught and pointed out an error early on—one that I'm glad didn't make it into the pages of this book.

On a visit to Harlingen in the Netherlands, I was able to go aboard the replica of William Barents's ship, which was under construction before and during the years I spent writing this book. Celestial navigation expert Dick Huges spent weeks chatting with me by video, teaching me the basics of sailing using only the sun and stars to determine one's location on the planet. Dick also coordinated a series of meetings in the Netherlands with experts working on projects related to Barents, to which he then spent a week driving me. Physicist Siebren van der Werf has explored the genesis of historical navigation tables and written a whole book on the

Nova Zembla effect, and he invited me into his home to discuss all of it. After years of work on shipwrecks and review of the drawings from the first account of Barents's voyages, Gerald de Weerdt developed plans to rebuild Barents's ship, and led the project to fruition, taking me aboard and answering questions for hours. Thanks also to Koos Westra, the mechanical engineer involved in rebuilding Barents's ship, whose small-scale replica of the vessel answered so many of my questions.

I'm likewise grateful for state museums and libraries. The archives of public institutions provided much of the material on which this book is based. Visits to the Rijksmuseum between 2015 and 2019 were particularly helpful. Actually examining the relics recovered from Barents's cabin with Jan de Hond was a gift. Seeing the castaways' buttons and shoes and hand tools made the crew come alive in my mind. The archives of the Netherlands' National Maritime Museum provided many sources and resources that helped me get a better sense of Dutch seafaring during the era.

The spine of this book is built from the first-person accounts of Gerrit de Veer and Jan Huygen van Linschoten. Countless visits to the Library of Congress, including a chance to photograph a four-hundred-year-old manuscript of van Linschoten's work, provided material for translation and a whole second viewpoint on Barents's first two voyages that had been missing from English-language accounts. As magnificent as any of these was the warm welcome and chance to enjoy the banya at the Russian Arctic National Park outpost on Cape Desire at the northern end of Nova Zembla.

I was fortunate enough to go on three Arctic expeditions during the course of researching and writing this book—all to regions visited by William Barents. The first, in January 2018, during polar night, was a dogsled expedition to the interior of Spitsbergen, thanks to the companionship and skills of Marcel Starinsky, Traci

Acknowledgments

Crippen, Sarah Marshall, Stina Stovring Andersen, and Lars Broens. Musher Blair Braverman was the inspiration for this trip, suggesting I look into kennels on Svalbard after she was generous enough to invite me to Wisconsin and teach me to dogsled in February 2017.

My next trip to Svalbard, whose existence Barents was the first to record, happened in the fall of 2018 as part of the Arctic Circle residency program, which takes participants along the western coast for weeks aboard a tall ship, and teaches interested passengers to haul lines, shift sails, and reckon a course. I spent as much time as possible up the mast to try to see the places Barents sailed through the eyes of those who sailed with him. On that voyage, Captain Mario Czok, First Mate Marijn Achterkamp, Second Mate Annet Achterkamp, Piet Litjens, Jana Maxovà, Janine Jungermann, and Alex Renes were indispensable. Just as crucial were expedition leader Sarah Gerats and guides Kristin Jæger Wexsahl, Åshild Rye, and Emma Hoette—who led our hikes, provided historical context, guarded us from polar bears, and sacrificed their own comfort and safety to ferry several people over a rising current. While in Longyearbyen, I was also able to meet with Alexander Hovland and visit the replica of William Barents's cabin, which had been re-created on Svalbard hundreds of miles from Nova Zembla.

The third expedition I went on was to the Russian Arctic, sailing out of Murmansk to Nova Zembla (Novaya Zemlya in Russian), where I visited the most important sites mentioned in Gerrit de Veer's account of William Barents's voyages. That trip stands as one of the happiest experiences of my life. The other passengers on that voyage—Marthe Larsen Haarr, Michael Pantalos, Alexey Neumoin, our cook Olga Chumachenko, and especially Tatiana Ponomareva, who had the thankless task of interpreting for me—sailed on with grace and fortitude for a week longer than we expected to remain in such close quarters. Thank you to Mike Chernobylsky and Victor Boyarsky at VICAAR for organizing the expedition, and to Natalia Krutikova, Alexandr Chichaev, and Maria Gavrilo for making it possible for me to sail aboard the *Alter Ego*. Thanks, too, to Vadim and Alexander, the park rangers at the Nova Zembla station during our stops at Cape Desire.

Thank you most of all to the crew on that expedition, each of whom was a wonder and a gift. Mikhail Tekuchev, the captain, let me try my hand at learning to sail and steer the boat from the start. First mate Andrey Ianushkevich, the ship's mechanical wizard, likewise helped me learn the ropes and performed so many kindnesses on sea and land that my attempts to keep thanking him became embarrassing for both of us. Evgeny Fershter's experience with polar bears and expertise on the Barents sites made each stop more productive than hours spent in any archive. Alexander Bogdanov's hard work, joy, and musical accompaniment in the face of every unexpected event (engine failure, mirage, pods of walruses) carried the day. It wouldn't be possible to travel with a more delightful group.

As always, I am beyond grateful to my family. Patti Pitzer, Terri Ellis, Peter Vergano, and Kathy Vergano helped out with my kids in ways large and small, taking up tasks that I ran out of time for or abandoned altogether while working on this book. My two children got used to my staying up long after they went to bed, and vanishing on expeditions or research trips. They've become proficient at reminding me of things I forget to take care of, and have turned into wonderful human beings, despite sometimes being left to forage for themselves. And my husband, Dan, continues to be the steady one—the one who's willing to hear out every wild idea, who embraces my eccentricities alongside whatever better traits I occasionally display, and who believes in me even when I lose faith in myself.

Notes

Chapter One: The Open Polar Sea

1 Diederick Wildeman, "Who Was William Barents?" Original, untranslated source: "Wie was Willem Barentsz? De rol van Barentsz tijdens de reizen naar het hoge noorden in 1594–1597," in Leo Akveld, Remmelt Daalder, Frits Loomeijer, Diederick Wildeman, eds., *Koersvast: Vijf eeuwen navigatie op zee: Een bundel opstellen aangeboden aan Willem Mörzer Bruyns bij zijn afscheid van het Nederlands Scheepvaartmuseum Amsterdam in 2005* (Zaltbommel: Aprilis, 2005), p. 218. Translated into English for the author in December 2018 by Robert Neugarten.

2 Ole Peter Grell, *Calvinist Exiles in Tudor and Stuart England* (London: Routledge, 1996), p. 4.

3 Richard Paping, "General Dutch Population development 1400–1850, cities and countryside," University of Groningen, 2014. https://www.rug.nl /research/portal/files/15865622/articlesardinie21sep2014.pdf.

4 Diodorus Siculus, *Library of History* Book II. 47. 1–6 (trans. C. H. Oldfather).

5 Pepijn Brandon, Sabine Go, and Wygren Verstegen, *Navigating History: Economy, Society, Knowledge, and Nature: Essays in Honour of Prof. Dr. C.A. Davids* (Amsterdam: International Institute of Social History, 2018), p. 133.

6 Thomas R. Rochon, *The Netherlands: Negotiating Sovereignty in an Interdependent World* (Boulder, CO: Westview Press, 1999), p. 237.

7 Jürgen G. Backhaus, *Navies and State Formation* (Vienna: LIT Verlag, 2012), p. 283.

8 This work of Pytheas only exists in fragments quoted by others. This quote comes from Strabo, as translated by Christina Horst Roseman in *Pytheas of Massalia: On the Ocean* (Chicago Ridge, IL: Ares, 1994), p. 125.

9 "Proof of a 2,000 kilometre polar trade route in volcanic glass dating back at least 8,000 years," *Siberian Times*, March 7, 2019.

10 One account says he began with only twenty-five ships, but in either case, he lost a staggering number of ships in transit.

11 Soren Thirslund, *Viking Navigation* (Roskilde: Viking Ship Museum, 2017), p. 11.

12 Siebren van der Werf, "History and critical analysis of fifteenth and sixteenth century nautical tables," *Journal for the History of Astronomy* 48, no.2 (May 2017):207–232.

13 Van der Werf, "History and critical analysis," p. 3.

14 "The Ship's Council on the Expedition of Pet and Jackman on July 27th, 1580," *Mariner's Mirror* 16, no. 4 (1930):411.

15 Jan Huygen van Linschoten, *Voyagie, ofte schip-vaert, van Ian Huyghen van Linschoten, van by Noorden om langes Noorvvegen de Noortcaep, Laplant, Vinlant, Ruslandt, de Vitte Zee, de custen van candenoes, Svvetenoes, Pitzora* (Amsterdam: Ian Evertss. Cloppenburg, 1624), preface, p. 3. Translated into English for the author in 2018 and 2019 by Tjitske Kummer.

Chapter Two: Off the Edge of the Map

1 Jan Huygen van Linschoten, *Voyagie, ofte schip-vaert, van Ian Huyghen van Linschoten, van by Noorden om langes Noorvvegen de Noortcaep, Laplant, Vinlant, Ruslandt, de Vitte Zee, de custen van candenoes, Svvetenoes, Pitzora*, preface, *Pitzora* (Amsterdam: Ian Evertss. Cloppenburg, 1624) p. 3. Translated into English for the author in 2018 and 2019 by Tjitske Kummer.

2 Ibid., preface, p. 2.

3 Michael Engelhard, *Ice Bear: The Cultural History of an Arctic Icon* (Seattle: University of Washington Press, 2017), p. 102.

4 Gerrit de Veer, *The Three Voyages of William Barents to the Arctic Regions* (London: Elibron Classics, 2005), p. 25.

5 Ernest Shackleton, *South: The Illustrated Story of Shackleton's Last Expedition: 1914–1917* (Minneapolis: Zenith Press, 2016), p. 99.

6 Princeton University Maps Library, catalogue information on Ferdinand Magellan. Pulled December 1, 2019. https://libweb5.princeton.edu/visual _materials/maps/websites/pacific/magellan/magellan.html.

Chapter Three: Death in the Arctic

1 Diederick Wildeman, "Who Was William Barents?" endnote 3. Original, untranslated source: "Wie was Willem Barentsz? De rol van Barentsz tijdens de reizen naar het hoge noorden in 1594–1597," in Leo Akveld, Remmelt Daalder, Frits Loomeijer, Diederick Wildeman, eds., *Koersvast: Vijf eeuwen navigatie op zee: Een bundel opstellen aangeboden aan Willem Mörzer Bruyns bij zijn afscheid van het Nederlands Scheepvaartmuseum Amsterdam in 2005* (Zaltbommel: Aprilis, 2005), p. 215. Translated into English for the author in December 2018 by Robert Neugarten.

2 Gerrit de Veer, *The Three Voyages of William Barents to the Arctic Regions* (London: Cambridge University Press, 2012), p. 60, footnote one.

Notes

3 Elaine Fantham, "Caesar and the Mutiny: Lucan's Reshaping of the Historical Tradition in De Bello Civili," *Classical Philology* 80, no. 2 (April 1985).

Chapter Four: Sailing for the Pole

1 Diederick Wildeman, "Who Was William Barents?," paragraph 10. Original, untranslated source: "Wie was Willem Barentsz? De rol van Barentsz tijdens de reizen naar het hoge noorden in 1594–1597," in Leo Akveld, Remmelt Daalder, Frits Loomeijer, Diederick Wildeman, eds., *Koersvast: Vijf eeuwen navigatie op zee: Een bundel opstellen aangeboden aan Willem Mörzer Bruyns bij zijn afscheid van het Nederlands Scheepvaartmuseum Amsterdam in 2005* (Zaltbommel: Aprilis, 2005), p. 215. Translated into English for the author in December 2018 by Robert Neugarten.

2 Ibid., endnote 5.

3 Fletcher Bassett, *Legends and Superstitions of the Sea and of Sailors in All Lands and at All Times* (London: S. Low, Marston, Searle & Rivington, 1885), p. 418.

4 Author interview with Peter J. Capelotti, January 2018.

Chapter Five: Castaways

1 William Dean Howells and Thomas Sergeant Perry, *Library of Universal Adventure by Sea and Land* (New York: Harper & Brothers, 1888), p. 23.

2 "Dr. Rae's Report," A letter from Rae to Charles Dickens, *Household Words* 10, no. 249 (December 30, 1854):458.

3 James S. Aber syllabus for History of Geology, Emporia State University. Pulled December 1, 2019. http://academic.emporia.edu/aberjame/histgeol/nansen/nansen.htm.

4 Susan Kaplan and Genevieve LeMoine, *Peary's Arctic Quest: Untold Stories from Robert E. Peary's North Pole Expeditions* (Lanham, MD: Down East Books: 2019), p. 24.

5 Peter J. Capelotti. *The Greatest Show in the Arctic: The American Exploration of Franz Josef Land, 1898–1905* (Norman: University of Oklahoma Press, 2016), p. 35.

6 Michael F. Robinson, *The Coldest Crucible: Arctic Exploration and American Culture* (Chicago: University of Chicago Press, 2006), p. 80.

7 Louwrens Hacquebord, "In Search of Het Behouden Huys," *Arctic* 48, no. 3 (September 1995):248.

Chapter Six: The Safe House

1 Svalbard Museum in Longyearbyen, Norway. Information pulled December 1, 2019. https://svalbardmuseum.no/en/kultur-og-historie/hvalfangst/.

2 Canadian Museum of History. Information pulled December 1, 2019. https://

www.historymuseum.ca/cmc/exhibitions/archeo/paleoesq/pea01eng
.html.

3 Iris Bruijn, *Ship's Surgeons of the Dutch East India Company: Commerce and the Progress of Medicine in the Eighteenth Century* (Leiden: Leiden University Press, 2009), p. 16.

4 Ibid., pp. 15–16.

5 Richard Unger, *Beer in the Middle Ages and the Renaissance* (Philadelphia: University of Pennsylvania Press, 2004), p. 130.

6 Sarah Bankhead, "Alcohol vs. Water: There is No Contest For 17th Century Sailor," Institute of Nautical Archaeology, March 13, 2017. Information pulled December 1, 2019. https://nauticalarch.org/alcohol-vs-water-there-is-no-contest-for-17th-century-sailors/.

7 Simon Worral, "A Nightmare Disease Haunted Ships in the Age of Discovery," *National Geographic*, January 2017, https://news.nationalgeographic.com/2017/01/scurvy-disease-discovery-jonathan-lamb/.

8 Jeremy Hugh Baron, "Sailors' scurvy before and after James Lind," *Nutrition Reviews* 67, no. 6 (2009):315–332.

9 Rachael Rettner, "How Does a Person Freeze to Death?" LiveScience, January 30, 2019. https://www.livescience.com/6008-person-freeze-death.html.

10 Peter Stark, "Frozen Alive," *Outside* magazine, March 7, 2016.

11 Siebren van der Werf, email exchange with the author, February 18, 2020.

12 Heinz Mehilhorn, *Encyclopedic Reference of Parasitology: Biology, Structure, Function* (Berlin: Springer, 2001), p. 289.

Chapter Seven: The King of Nova Zembla

1 Anke A. Van Wagenberg-Ter Hoeven, "The Celebration of Twelfth Night in Netherlandish Art." *Simiolus: Netherlands Quarterly for the History of Art* 22, no. ½ (1993):65–96. doi:10.2307/3780806.

2 Hjalmar Johansen. *With Nansen in the North: A Record of the Fram Expedition in 1893–96* (Ann Arbor: University of Michigan Library, 1899), p. 120.

Chapter Eight: The Midnight Sun and the False Dawn

1 Siebren van der Werf, *Het Nova Zembla Verschijnsel: Geschiedenis van een Luchtspiegeling* (Historische Uitgeverij: 2011), and W. H. Lehn and I. I. Schroeder, "Polar Mirages as Aids to Norse Navigation," *Polarforschung* 49, no. 2 (1979):173–187.

2 Priscilla Clarkson, "The Effect of Exercise and Heat on Vitamin Requirements," *Nutritional Needs in Hot Environments Applications for Military Personnel in Field Operations,* Institute of Medicine, US Committee on Military Nutrition (Washington, DC: National Academies Press, 1993).

Notes

3 John McCannon, *Red Arctic* (New York: Oxford University Press, 1998), p. 48.

Chapter Nine: Escape

1 "Warming to Cap Art," *The Journal*, August 15, 2006.

2 Diederick Wildemann, "Who Was William Barents?" Original, untranslated source: "Wie was Willem Barentsz? De rol van Barentsz tijdens de reizen naar het hoge noorden in 1594–1597," in Leo Akveld, Remmelt Daalder, Frits Loomeijer, Diederick Wildeman, eds., *Koersvast: Vijf eeuwen navigatie op zee: Een bundel opstellen aangeboden aan Willem Mörzer Bruyns bij zijn afscheid van het Nederlands Scheepvaartmuseum Amsterdam in 2005* (Zaltbommel: Aprilis, 2005), p. 219. Translated into English for the author in December 2018 by Robert Neugarten.

3 J. H.G. Gawronski and P. V. Boyarsky, eds., *Northbound with Barents: Russian-Dutch Integrated Archaeological Research on the Archipelago Novaya Zemlya* (Amsterdam: Stichting Olivier van Noort, 1997), p. 92.

Coda: The Shores of Nova Zembla

1 T. C. W. Blanning, *The Pursuit of Glory: Europe, 1648–1815* (New York: Viking, 2007), p. 96.

2 Diederick Wildeman, "Who Was William Barents?" Original, untranslated source: "Wie was Willem Barentsz? De rol van Barentsz tijdens de reizen naar het hoge noorden in 1594–1597," in Leo Akveld, Remmelt Daalder, Frits Loomeijer, Diederick Wildeman, eds., *Koersvast: Vijf eeuwen navigatie op zee: Een bundel opstellen aangeboden aan Willem Mörzer Bruyns bij zijn afscheid van het Nederlands Scheepvaartmuseum Amsterdam in 2005* (Zaltbommel: Aprilis, 2005), p. 217. Translated into English for the author in December 2018 by Robert Neugarten.

3 Svalbard Museum in Longyearbyen, Norway. Section on whaling history in permanent exhibition.

Index

Page numbers in *italics* refer to maps and illustrations.

Act of Abjuration, 4
Admiralty Island, 37, 113, 238
adzes, 135, 150
Africa, 68, 220
   northern, 13
   sailing around, 96, 219
   southern, 19
airplanes, 176
airships, 149
   hydrogen, 129
Alaska, 44
Aldebaran (star), 180
Alps, 130
Alva, Third Duke of, 4
American explorers, 129, 131
amnesia, 164
amphibolite, 20
Amsterdam, 6–7, 10, 16, 19, 22, 24, 27,
     39, 60, 65, 96, 174, 258, 260
   canals of, *3*
   Court of the Admiralty in, 260
   harbor of, *9*, 26, 28, 63, 69
Amundsen, Roald, 38, 148–49, 176, 219, 223
anchors, 53, 78, 105, 107
   bower, 137
   grappling, 86
   kedge, 84, 117, 119
   spare, 143

Andries, Claes, 132, 211
   death of, 217, 259
Antarctica, 44
Antwerp, 4, 5
Arabs, 17, 107
Arawak people, 45
Archangel, 79
Arctic Circle, 17, 31, 52, 263
Arctic hares, 86, 87
Arctic loons, 48
Arctic poppies, 130
Arctic regions, 19, 29, 31, 41, 51, 99,
     113, 128
   Greek and Roman knowledge of,
     12–13, 16–17
   survival in, 130–31
   training for expeditions in, 128
   vegetation in, 48, 130, 138
Aristotle, 100
Armadillo (Boyd), 262
arquebuses, 103
ascorbic acid (vitamin C), 159, 160,
     214, 243
Asia, 10, 13, 21, 29
   eastern coast of, 57
astrolabe, 17–18
astrology, 167
astronomy, 17–18, 107, 185

Index

Asvaldsson, Thorvald, 14
Atlantic Ocean, 12, 16
    crossing of, 19–20, 35, 126
    North, 14
aurora borealis, 93
awls, 150

Babylonians, 18
bacon, 133, 211, 250
Bacon, Francis, 107
Bali, 219
Baltic Sea, 3
Baltic states, 12
barber-surgeons, 132, 153–54
Barents, William, 1–41, 129–30, 260
    birth of, 2, 218
    childhood and adolescence of, 2–3, 4
    death of, 3, 108, 217, 220, 222–24, 259
    failures of, 218, 220, 221, 223
    fame of, 3, 218–19, 222–24, 263
    first Arctic voyage of, 1–2, 5–7, 9–12, 16, 17–22, 24–28, 31, 33–39, 41–42, 44–45, 47–48, 51–52, 55, 59–63, 66, 67, 104, 110
    illness of, 211, 216, 217
    knowledge and skills of, 25, 132, 225
    mapmaking of, 3, 217, 218, 220, 245–46
    Mediterranean voyages of, 3, 12
    navigation of, 130, 180, 215, 217, 225, 245–46, 263
    overwintering plan of, 130, 131, 221
    physical appearance of, 3
    planned exploration voyages of, 126
    provisions carried at sea by, 24, 44
    relationship of Rijp and, 100–103, 109–10
    second Arctic voyage of, 65–68, 71, 73, 81–84, 86, 88–89, 92–96, 110, 130, 131, 180, 219, 240
    third Arctic voyage of, 97, 99–101, 103–10, 112, 115, 118–24, 131–32, 148–49, 173–75, 177, 179, 182–85, 214–15, 220
    training and education of, 2, 3
    wife and children of, 25
Barents Sea, 218
barges, 91
barrels, 54, 103, 133, 139, 140, 142–43
    chimney made from, see chimney, at Safe House
    makeshift sauna from, 153, 158, 195
Bashkortostan, 3
baths, 153
"Battle of the Books" (Swift), 261
bears, 31, 50, 136–37
Beechey Island, 126
beer, 24, 133, 150, 155, 161, 257
    spruce, 140
Beynen, L. R. Koolemans, 221
Beyren Eylandt (Bear Island), 104, 106–9, 115, 255, 257
Bibles, 175–76
Black Death, 153, 205
blisters, 165
blizzards, 123, 152, 258, 192
blood, 136
    circulation of, 25
boat hooks, 114
boats, aluminum, 129
bows and arrows, 56, 82
Boyarsky, Pyotr, 224
Boyd, William, 261–62
Brahe, Tycho, 19
brandy, 133
bread, 123, 133, 139, 155, 182, 232, 244, 246, 249, 250–51, 253
Brielle, 11
Britain, 12, 13, 126, 142
Brontë, Charlotte, 261
Brunel, Olivier, 21, 47, 59
buckets, 62

Bulgaria, 3
Burma, 67

cabins, building of, 132, 135–38, 140–41
Cabot, John, 19–20
Caesar, Julius, 90
calculus, 25
California, 3
Canada, 20, 126, 127
cannibalism, 127, 222
cannons, 132
    wheeled, 98
Cape Comfort, 227
Cape Desire, 119, 120, 214
Capelotti, P. J., 105, 129
Cape of Good Hope, 29
Cape of the Eleven Thousand Virgins,
    19
Cape of the Idols, 75, 78
Capo Baxo, 36
Capo Negro, 36–37, 238
Carcass, HMS, 142
carpenters, 132, 135, 183–84
cartography, 19
castaways, 125–45
Cathay, 10, 20, 173
central heating, 126
chaffinches, 52
cheeses, 211
chimney, of Safe House, 137, 147, 149,
    161, 163, 139, 172–73, 181, 184,
    189, 191, 193, 202, 209
China, 15, 21, 39, 57, 59, 63, 131, 173, 241
    eastern shores of, 67
    inventors in, 18
    search for a northern trade route
        to, 1–2, 6, 7, 10, 19, 20, 27, 29,
        60, 62, 65, 96, 108, 218–19, 223
    search for a southern trade route
        to, 2, 28
    search for a western trade route
        to, 20

chisels, 150
Cicero, 100
climate, adapting to, 130
clockmakers, 108
clouds, 130, 182, 183
coal, 126, 148, 163–64, 181–82
cod, 61
Cold Sea, 50
Columbus, Bartholomew, 45
Columbus, Christopher, 18, 19, 108, 218
    final voyage of, 45
    mutiny against, 45
Company of Cathay, 20
compasses, 15, 16, 80, 107–8, 109, 126,
    247, 248–49, 251
construction methods, of cabin,
    135–37
Cook, James, 153
cooks, 132, 253
Copernicus, Nicolaus, 19, 106
crayfish, 57
Cross Cape, 78
Cross Island, 39, 236, 238
cross-staff, 35
crowbars, 121
Cruyshoek, 91
currents, 120, 122, 138
cutlasses, 43, 88

Danish fishermen, 164
Declaration of Independence,
    American, 5
deer, hunting of, 125
Delft, 259
Delgoy Island, 48
Denmark, 260
Descartes, René, 107
Dickens, Charles, 127
disease
    contagious, 204, 205, 219
    germ theory of, 25, 155
    see also scurvy

Index

dogs, 125, 145

driftwood, 51, 54, 106, 133, 151–52, 166

drowning, 84

ducks, 50, 61

Dutch language, 2, 17, 21, 263

Dutch merchants, 47, 65, 66, 68, 97

Dutch national anthem, 221

Dutch Republic:
  birth of, 16, 21
  map of, 8
  see also Netherlands

Earth, 1
  first circumnavigation of, 19, 90
  gap between magnetic and polar north of, 108
  shape and demarcations of, 16
  Southern Hemisphere of, 68
  Western Hemisphere of, 18

East Indies, 219, 220, 269

eggs, 103, 106, 110, 119, 216, 238, 239

Egmont, Count of, 4

Egyptians, 18

Elcano, Juan Sebastián, 19

electricity, 25

elephants, 67

Elizabeth I, Queen of England, 11, 20

Endurance, The (ship), 44

England, 19–20, 126, 127

Enkhuizen, 10, 22–23, 28, 63, 66, 68, 73, 95, 96

Epiphany, feast of the, 174

equator, 16, 17, 100

Erebus, The (ship), 126

Eriksson, Leif, 14

Erik the Red, 14

"Essay on Man, An" (Pope), 261

Europe, 2, 10, 15, 16, 18, 20, 23, 106
  Renaissance, 107
  Scientific Revolution in, 19, 106–7
  Western, 3, 12

falcons, 61, 65

fall equinox, 124, 134

Far East, 19, 20, 24, 59, 66, 173
  searching for a northwest passage to, 20, 108, 126
  slave trade in, 56

Far North, 130–31, 261

Faroe Islands, 15

Finnei tribe, see Sami people

Finns, 31–32

firewood, 151, 154, 165, 172, 179, 234

fish, 51, 133, 172, 247, 252

fishing, 128, 252

fishing nets, 32–33, 76

Flanders, 22

fluyt, 9–10

fog, 35, 37, 47, 49, 51, 58, 59, 60, 70, 85, 86, 115, 229, 248

forests, 52, 61

foxes, 77, 160–61, 182, 188–89
  skinning and cooking of, 150–51, 168
  trapping and killing of, 150, 151, 152, 155, 156, 158–61, 164–65, 166, 168, 181

Franklin, John, 126–28, 130, 175, 222

Franz Josef Land, 148, 176

French language, 263

Frobisher, Martin, 20, 42, 221

furs, 77, 82, 130, 131, 161, 166, 260

Galileo, 19

Gama, Vasco da, 218

Geats tribe, 13

geese, 48, 50, 61
  barnacle, 106, 110, 263
  eggs of, 106, 110
  killing of, 106
  nesting of, 106, 110

Gemma, Cornelius, 167

George Henry (ship), 175

Germans, 131

glaciers, 36, 49, 55, 130
gnomon (sundial), 15
Goa, 67
Goa, Archbishop of, 23
gold, 42, 218
    mining of, 20
graves, 88, 135, 184
gravity, 25
Greek Orthodox Christians, 80
Greeks, ancient, 6, 12, 13, 16–18, 77
Greenland, 14, 15, 101, 108, 125, 128,
    129
    eastern coast of, 104
    northern, 149
Greyhound (ship), 68, 71, 81, 92
Griffin (ship), 76, 78, 81, 84, 87, 91, 92
    cargo carried on, 68
guillemots, 36
gunpowder, 108–9, 142, 193
guns, 43, 83, 87, 88, 89, 103, 108–9, 114,
    115, 123, 125, 133, 139, 142, 150,
    193, 202, 203, 234

Haarlem, 4, 23, 95
haddock, 61
Hague, 65, 259
    Royal Museum at, 99–100
hail, 52, 92–93, 206
halberds, 24, 103–4, 144, 145, 150
Hall, Charles Francis, 129, 131, 175–76
ham, 133
Hamburg Zoo, 38
hammers and nails, 136
Hampton Roads, Va., 205
Haroun and the Sea of Stories (Rushdie),
    261
harpoons, 50
Hasselaer, Peter, 260
hatchets, 24, 43, 103, 152
Hayes, Isaac Israel, 129
Haytham, Hasan Ibn al-, 107
hazing rituals, 100

Heemskerck, Jacob van, 221, 223–24
    and Barents's death, 217
    death of, 270
    descriptive letter of, 209–10, 211
    as East Indies fleet commander, 269
    in escape from Nova Zembla by
        small boat, 214, 233, 239–41,
        243–45, 252–53, 255–59
    in meetings with Russian sailors,
        240–41, 247, 249–51
    at Nova Zembla Safe House,
        157–58, 162, 165, 174–75,
        182–84, 187, 191, 202
    in planning for escape from Nova
        Zembla, 196–98, 200–201, 209
    polar bear attacks and, 113–14, 144,
        193
    Rijp's letter received by, 256–57
    in second Arctic expedition, 68, 180
    as ship captain in third Arctic
        expedition, 97, 110, 132, 220
    sun's reappearance spotted by,
        182–84
    surviving crew listed by, 211
    third expedition cargo supervised
        by, 132, 157
    Twelfth Night celebration approved
        by, 175
Henry VIII, King of England, 6, 19
Hernam, 15
Het Behouden Huys, *see* Safe House
*History or Description of the Great Empire
    of China, The,* 150
Hjaltland, 14, 15
Hope (ship), 68, 71, 73, 78, 81, 83, 84,
    86, 88–89, 92
horses, 83
hourglasses, 101, 151, 185
Houtman, Cornelis de, 219
Hudson, Henry, 90, 221
hunger, 127, 141, 241, 244, 246
hunting, 86, 125, 128

Index

hurricanes, 104
Huysduynen, 63
Hvarf, 15

Iberian Peninsula, 13
ice, 42, 47, 69, 73–87
  blocks of, 134, 239
  canyons of, 112
  crystals of, 99
  destructiveness of, 44–45, 47, 49, 53,
    54, 57, 67, 77, 83–86
  drift, 117, 131
  flat tables of, 41, 48, 80
  floating, 45, 48, 52–53, 78, 80–83,
    85, 112, 125
  fracturing of, 117, 122
  impassable, 74–76, 78–81, 84, 91, 96,
    101, 107, 112, 114
  melting of, 51, 79
  pancake, 41, 104
  shards of, 239
  ships trapped by, 81, 126–27, 241
  struggles of survivors in small
    boats through, 234–40, 244
  third Barents expedition trapped by,
    121–24, 137–39
  towers of, 41, 117, 239
  walking on, 81
icebergs, 41, 43, 45, 48, 49, 57, 75–76,
  80, 112
  breaking up of, 51, 74
  calved, 120
  collision with, 53, 126, 215
  floating, 52–53, 83
  grounded, 115, 117–18, 119
  size of, 51, 54, 73–74, 85–86, 101
Ice Harbor, 119–21, 131, 136, 141, 149,
  164, 213, 226, 245
  map of, *111*
ice-knees, 123, 124, 136
Iceland, 12, 15
  Hornstrandir settlement on, 14

Ice Point, 42–43, 47, 115, 214, 215
India, 6, 23
Indies, 18, 221
influenza, 204
Inughuit, 129
Inuit, 20, 127, 128, 129, 149, 175
Inuit language, 128
iron, forged, 151
Iron Pig (ship), 73
Italian language, 263
Italy, 3
ivory, 43, 57

Jackman, Charles, 20–21, 221
Jamaica, 45
Jane Eyre (Brontë), 261
Japan, 57
  tea drinking in, 67
Java, 219
Johansen, Hjalmar, 148, 176, 223
John of Harlem, 132, 259

Kanin Nos, 245, 247, 248, 249, 250, 251
Kara Sea, 57, 221
kedging, 84
keelhauling, 77–78, 90
Kildin Island, 28, 31–35, 37, 44, 48, 49,
  62, 92, 253, 256, 258
  inhabitants of, 51, 56
  maps of, *30, 34, 242, 254*
Kinbote, Charles (char.), 261
knarr (oceangoing ships), 14
Kola, 255–56, 258
  map of, *254*
Kola River, 259
Kostin Shar, 47, 240
Kravchenko, Dmitri, 224
kvas, 255

ladyas (Russian ships), 47, 49, 76
lances, 125
Landnámabók (Vilgerdarson), 15

Index

Langenes, 36
lanterns, 188
Lapland, 226
Lapplanders, 128
Latin language, 58, 62, 263
lichens, 36, 42, 130
Linschoten, Jan Huygen van, 90, 97
    diary of, 51
    first northern voyage of, 23–24,
        27–29, 32, 51–52, 54–58, 60,
        62–63, 75, 77
    intelligence collected and published
        by, 67–68
    merchants' interests represented by,
        65, 66
    second northern voyage by, 65–68,
        92–94
    wooden idol stolen by, 58, 77
liquor, 257
    see also beer; wine
Lisbon, 17
Little Ice Point, 117
Lofoten Islands, 31
Lombsbay, 36
Longyearbyen, 3
Low Countries, 4, 5
    see also Netherlands
lumber, 134–35, 137, 138
lunar declination tables, 45

Magellan, Ferdinand, 218
    first circumnavigation of the Earth
        by, 19, 90
    mutiny against, 45, 90
malaria, 204
mapmaking, 3, 6, 19, 22, 96, 217, 218,
    219, 220
Marseille, 12, 263
Massalia, 12, 13
Matfloe Island, 48
mathematics, 18
Maurice, Prince of Orange, 24, 43, 65

Maurice Island, 61
Mealhaven, 48, 240
meats, 251, 256, 257
    fox, 157, 159–61
    polar bear, 105, 106, 196, 203
    salted, 133, 159
Mediterranean Sea, 3, 12, 23, 217
Mercator, Gerardus, 22
Mercury (Mercurius; ship), 10, 23–24, 28,
    29, 32, 33, 45, 48, 49, 54, 63, 67
    running aground of, 61–62
Mercury River, 59
mermaids, 99–100
Meta Incognita, 20
Meteorology (Aristotle), 100
Meuse River, 259
Michelet, Jules, 153
microscopes, 107
Middle Ages, 153
mirages, 58, 185, 186, 263
monkeys, 56
moon, 17, 154, 167
    eclipse of, 45
    tides influenced by, 12–13
Morocco, 3
Moscow, 66
Moscow, Grand Duke of, 32, 56, 67, 258
mosses, 130
Moucheron, Balthazar de, 21, 22
Muscovy Company, 20
Muslims, 18
mutiny, 29, 45, 89–90, 131–32
    punishment for, 90, 96, 131

Nabokov, Vladimir, 261
Nansen, Fridtjof, 38, 128–29, 176, 185,
    223
Nassau, 40
Native Americans, 127
Nay, Cornelis
    first northern voyage of, 28, 33, 48,
        50, 54, 60, 62–63

Nay, Cornelis (*cont.*)
  second northern voyage of, 68, 75, 76, 81, 83–84, 89–90
Nelson, Horatio, 142
Nenets herders, 47, 55–56, 80, 82–84, 130, 131
Neptune, god of the sea, 100
Netherlands, 7–12, 56, 173, 245, 258
  admiralty boards of, 11
  art and culture of, 2, 7, 12
  cold winters in, 130
  in 1594 war with Spain, 1, 2, 3, 4–5, 33
  independence sought by, 5
  Portuguese rivalry with, 23
  provinces of, 10, 12, 22, 28, 63, 65, 66, 68, 96
  provincial council of, 96
  return of the surviving Barents crew to, 259–60
  shipbuilding in, 2, 7–10
  slavery in, 2
  States General parliament of, 33, 58, 65, 76
  trade between Russia and, 21
  as world's leading economic and naval power, 2, 100
  Zeeland province of, 10, 22, 28, 63, 66, 68, 96
New Description and Atlas of the Mediterranean Sea (Barents), 3, 23, 217
New Holland, 61
New North Sea, 57
Nobile, Umberto, 223
Nordenskiöld, Nils Adolf Erik, 218–19
Nordkapp, 31, 70
Nordkinn, 31, 73, 93
Norge (airship), 149, 223
Norsemen, 128
North America, 15, 19–20, 108, 128
  Arctic, 108

Dorset culture in, 149
Hudson's search for a northern passage over, 90
remains of Viking settlements in, 14
North Pole, 16, 19, 28, 39, 42, 96, 97, 118, 126, 129, 131, 222, 223
  concept of temperate climate at, 6, 22
  maps of, *108*
North Sea, 7, 10, 28, 29, 35, 62, 164
North Star, 15
Norway, 6, 15, 21, 24, 28, 226
  coast of, 31, 42, 62, 70, 96, 112
  northern, 128
  Russian border with, 28
Norwegians, 217
Nova Zembla, 1–2, 21, 22, 24, 27–28, 33, 42, 44, 45, 49, 55, 62–63, 67, 110, 245, 261
  castaways' harrowing journey home in small boats from, 225–60, 263
  coast of, 36, 39–40, 41, 47, 48, 74, 113, 131, 224
  eastern, 96, 240
  Goose Land portion of, 47
  Ice Harbor at, 119–21, 131, 136, 141, 149, 164, 213, 226, 245
  Ice Point at, 42–43, 47, 115, 214
  maps of, *46, 111, 212, 242*
  northern, 40, 43, 51, 66, 89, 96, 214
  planning of escape from, 181, 196–216
  plans to overwinter on, 124
  Safe House at, *see* Safe House
  sailors stranded on, 131
  southern, 47, 79, 120
  third Barents expedition trapped by ice at, 121–24, 137–39
Nunavut, 20, 126

oars, 91–92, 247
On the Ocean (Pytheas), 12–13

Index

On the Revolutions of the Heavenly Spheres (Copernicus), 19
Orange Islands, 43, 47, 118, 214, 266, 267
Orion (constellation), 165
Orkney Islands, 12, 261

Pacific Ocean, 19, 126, 220
 South, 219
Paris, Viking conquest of, 13
Parry, William, 108
Peary, Robert, 129, 148, 223
Pechora River, 21, 51
Persia, 18
Pet, Arthur, 20–21, 221
Philadelphia Inquirer, 129
Philippines, 19
pilots, 126, 132
pinnace, 66, 70
piracy, 11, 29, 66, 77
plague, 204
Plancius, Petrus, 3, 10, 22, 27, 66–67, 97
 mapmaking of, 96, 219
planets, 17
plumb lines, 14, 35, 41, 58
polar bears, 13, 25, 37–39, 65, 112–18,
  140–42, 152–53, 180–83
 attacks by, 143–45, 221–22, 223–24
 chasing of, 118
 freezing of, 133–34, 139
 killing of, 88, 103–4, 105, 108–9,
  114–15, 118, 133–34, 139, 141,
  187–88
 meat of, 105, 106, 196, 203
 natural habitat of, 141
 pelts of, 107
 personalities of, 141
 sailors killed by, 87–88, 89
 skinning of, 88, 104, 105, 114–15,
  118, 130, 188
 sleeping, 118
Polaris expedition, 131
polar night, 180, 183, 185–86

Pomor (Russian sailors), 21, 36, 39, 67,
 221
Pope, Alexander, 261
Porras, Diego de, 45
Porras, Francisco de, 45
Portugal, Portuguese, 3, 5, 17, 67, 219
 Eastern trade empire of, 93–94
 exploratory voyages of, 68
 rivalry between Netherlands and, 23
prayer, 93, 122–23, 131, 182, 184, 185
printing press, 17
Protestantism, 5, 11, 19, 153
pumps, 62, 70
Purmer, Lake, 135
Purmerend, 135
pyroxenite, 20
Pytheas, 12–13, 16, 263

Rae, John, 127
rain, 61, 62, 69, 70, 81, 132, 206, 243,
 248, 252, 256
rainbows, 99, 100, 107
rats, 205
reindeer, 31–32, 52, 55, 56, 83, 131
 antlers of, 54
 skins of, 77, 82
reindeer bridles, 76
reindeer herders, 47, 55–56, 67, 78, 80,
 82–84, 130, 131
Rembrandt, 2
Riffenburgh, Beau, 222
Rijp, Jan Cornelis, 68, 97, 99, 131, 260
 Barents's relationship with,
  100–103, 109–10, 221
 in meeting with Nova Zembla
  survivors, 257–59
rock crystals, 58, 87
Roman Catholic Church, 4, 174
Roman writers, 13
Rotterdam, 11, 66, 91, 96
rowboat, 91, 103–4, 105, 106, 120, 122,
 202, 203, 205, 207, 208, 210

rowboat (*cont.*)
　　in escape from Nova Zembla, 228,
　　　231–32, 243, 244–45, 248–49,
　　　250–51, 253, 255–56, 258–59
Rushdie, Salman, 261
Russia, 1, 6, 20, 21, 24, 31, 49, 84, 221
　　coast of, 226, 245
　　Norwegian border with, 28
　　trade between Netherlands and, 21
Russian language, 28, 55, 61
Russians, 29, 31–33, 36, 39, 47–52, 55,
　　56, 57, 67, 77, 79–80, 217, 240–41,
　　244–45, 247–53, 255–58

Safe House (Nova Zembla wooden
　　cabin), 132, 135–38, 140–41,
　　147–69, 171–73, 260
　　abandonment of, 213–14, 216
　　deaths at, 183–84
　　discovery of remains of, 213
　　holiday celebrations at, 171, 174–75,
　　　177, 179
Salutation, The (ship), 125, 148
Sami people, 13, 31–32, 47, 56, 128,
　　131, 252, 255, 257
sandbars, 60
saw, windmill-driven, 9
Scandinavia, 10, 16, 217, 259
　　coastal tribes of, 13
scientific method, 107
Scotland, 261
Scott, Robert, 223
screw-pumps, 62
scurvy, 159–61, 165, 181, 183, 219
　　cause and treatment of, 25, 160–61,
　　　214, 241, 243, 255
　　symptoms of, 93, 159, 244
scute, 143, 201–2, 205–6, 207, 208, 209
　　in escape from Nova Zembla, 228,
　　　229, 231–33, 235, 243, 244–45,
　　　247–48, 249, 250, 251, 252–53,
　　　255–56, 258–59

Sea Beggars, 11
sea birds, 52, 61
seagulls, 52, 103, 104
　　eggs of, 103
seals, 48, 141, 203
Seine River, 13
Sephardic Jews, 5
Shackleton, Ernest, 44
shipbuilding, 2, 7–10, 98
ships
　　ballast on, 91, 105
　　bomb, 126
　　bunk beds on, 98
　　captain's quarters on, 98
　　cargo carried on, 68, 75, 97, 98, 132,
　　　150
　　collision between, 89
　　cook's stoves on, 98, 134
　　damaged, 70–73, 95, 97, 136
　　danger of underwater blockage to,
　　　91, 95
　　fire on board, 93
　　food and drink carried on, 133, 139,
　　　140–41, 143
　　hulls of, 117, 123, 126
　　libraries aboard, 126, 130
　　listing of, 123, 142
　　main decks of, 98
　　open toilets on, 98
　　orlop decks of, 29, 98
　　poop decks of, 138
　　portholes of, 98
　　rescue, 125–26
　　running aground of, 131
　　sails and masts of, 53, 59, 60, 71, 74,
　　　84, 85, 98, 120–21, 123, 213, 243
　　steam-powered, 126
　　tillers and rudders of, 119–20, 122
　　vanishing of, 125–28
　　whaling, 44
　　yachts carried on, 68, 76, 78–80, 82,
　　　84, 87, 88, 92, 94

ship's biscuits, 133, 155, 182
  see also bread
ship's clock, 151, 161–62
shipwrecks, 127
Siberia, 67, 138
Siberians, 13
skis, 128
slate, 152
slavery, 2, 56, 220
sleds, 55, 76, 77, 83, 130, 134, 135,
        143–45, 165–66
  aluminum, 129
smallpox, 204, 205
smoke, 139, 149
snow, 58, 92, 93, 113, 122, 132, 133,
        137, 138, 152, 161, 192, 206
  driving, 115, 134, 143
  melting, 54, 156, 157, 214
  mountains of, 103
  piling up of, 143, 158, 167, 168–69,
        174, 187, 189, 201, 234
  shoveling of, 140, 165, 167, 168
snowshoes, 128
South America, 19, 45
South Pole, 16
Spain, 11, 23
  exploratory voyages sponsored by,
        18
  in 1594 war with Netherlands, 1, 2,
        3, 4–5, 33
Spanish Fury, 4, 11
Spice Islands, search for a southern
        route to, 45
spices, 2, 219
Spitsbergen, 129, 131, 226, 257, 260,
        261
  coast of, 105
  map of, *102*
  Ny-Ålesund harbor at, 149
spoonwort, 243–44
Stappane, 31
stars, 167, 224

States Island, 58, 85–86, 88–90
steam engines, 126, 130
storms, 53, 54, 57, 62, 69–71, 78, 81–82,
        84, 91, 92–93, 108, 119–20, 122
  in escape from Nova Zembla, 243,
        248, 256
  at Nova Zembla, 137, 140, 152, 158,
        159, 166–67, 168, 190–91, 197–98
Strait of Gibraltar, 3, 12
Strait of Nassau, 53
summer solstice, 101, 110
sun, 15, 29, 47, 58, 73
  blocking of, 51
  dimming of, 134
  gazing at, *35*, 224
  midnight, 12, 32, 99, 184–86, 214
  parhelia phenomenon of, 99–100
  positions of, 17, 18, 37, 41, 42, 52,
        75, 104, 107, 134–35, 167
  rising and setting of, 52, 152, 251
sun dogs, 99
sunstones, 15
superstition, 99–100, 106, 167
swallows, 52
Swan River, 59
swans, 49, 61, 101
Swedish tribe, 13
Swift, Jonathan, 261

Tables of Toledo, 17
tapeworms, 167
Taqulittuq, 175–76
tarpaulins, 115, 207, 216
Tartary Coast, 67
Taurus constellation, 180
telescopes, 25, 107
Terror, The (ship), 126
Terschelling Island, 218
Texel Island, 28, 63, 69, 70, 73
thirst, 127, 227
Thorne, Robert, 19
Thule, 12–13, 149, 156

Index

tides, 13, 98, 130
tinderboxes, 125
Tollens, Hendrik, 221
treadmills, 62
trees, 132, 135, 246
trigonometry, 17, 18
Trondheim, 31
Tropic of Cancer, 17
Tropic of Capricorn, 17
trumpets, 57, 83
Twelfth Night, 174, 175, 177, 250
Twenty Thousand Leagues Under the Sea (Verne), 261

Ukraine, 13
unconsciousness, 164, 217
urit frigus (the cold burns), 58
Ursa Minor, 17

Vaigach Island, 21, 22, 24, 33, 48–53, 55, 73–77, 79, 82, 91, 94, 100–101, 226, 243, 276
    coast of, 51–53, 55, 57, 59, 60, 61, 67, 76, 80, 96
    north shore of, 53, 55
    safe harbor on, 58
    southern part of, 55, 75
    vegetation on, 52
Vaigach Strait, 21, 22, 24, 50, 52, 55, 56, 57, 62, 63, 66, 75, 78–80, 84, 240
    map of, 72
Valentyn, Mr., 100
Veer, Gerrit de:
    on Barents as expedition leader, 224
    and death of Barents, 217
    in escape from Nova Zembla, 213–17, 226, 228, 239–48, 252–53
    expedition journals of, 68, 82, 90, 106, 124, 132, 154, 158, 171, 186, 203, 213, 262–63, 269
    illness of, 252

    and near-disaster from cabin smoke, 163
    polar bear attacks and, 144, 208, 223
    in second Arctic expedition, 68, 82, 90, 180
    and sun's reappearance, 182–83, 185
vegetables, 133
Vermeer, Johannes, 2
Verne, Jules, 261
Victoria Strait, 126
Vikings, 35, 128
    clinker-style shipbuilding of, 14
    Danish, 13
    as navigators, 14–16
    northern exploration of, 13–16
    settlements of, 14, 16
vitamin C (ascorbic acid), 159, 160, 214, 243
Vilgerdarson, Flóki, 15
Vlie barrier islands, 98–99
Vlieland, 63
Vos, Hans, 132, 153

walruses, 43–44, 45–47, 50, 52, 57–58, 238
    carcasses of, 54, 65, 246
    skins of, 76, 79
    tusks of, 43–44, 57–58, 107
Wardhuys, 62, 92, 253, 255, 259
Warm Sea, 50
water
    desalinization of, 126, 127
    fresh, 124, 214
weather conditions, 40–41, 53, 54, 57, 61, 62, 69–71, 75, 77–79, 81–82, 85, 87, 94, 101, 115, 119, 123, 134, 135, 139, 172, 186–87, 246–47
Wellman, Walter, 129
Werf, Siebren van der, 165
whale oil, 76, 77, 241–42

Index

whales, 15, 50, 59, 70, 85
    dead, 91, 95, 104
    hunting for, 128
    jawbones of, 95
whaling ships, 44
White Sea, 32, 73, 79, 245, 247–48, 250, 251
whiting, 61
Wildeman, Diederick, 3, 224
William I, Prince of Orange, 4, 37, 43
    assassination of, 11
    piracy authorized by, 11
Williams Island, 37, 39
Willoughby, Hugh, 221
wind, 40, 41–42, 43, 44, 52, 59–62,
    69–71, 84–85, 91, 98, 104, 105,
    109, 112, 115, 118, 123
    in escape from Nova Zembla, 213,
      227–29, 235, 238–39, 246,
      248–51, 255, 258
    at Nova Zembla, 126, 130, 136, 143,
      162, 172–73, 189, 193, 201, 207–8

wine, 133, 139, 171, 177, 211,
    257
winter clothes, 93, 134
wolves, 31
wooden idols, 53–54, 55, 58, 75, 77,
    80, 83
woodworking skills, 132

yachts, 97
    carried on ships, 68, 76, 78–80, 82,
      84, 87, 88, 92, 94
yellow fever, 204, 205
Young Men's Christian Union of
    Cincinnati, 176

Zacuto, Abraham, 17
Zhukov Island, 13
Zuiderzee, 7, 10, 37, 63, 218, 259
Zutphen, 4
Zwaan (Swan; ship), 10, 24, 28, 33, 45,
    48, 49, 54, 61, 63, 67

About the Author

Andrea Pitzer loves to unearth lost or forgotten history. Her journalism has appeared in newspapers and magazines across the country. Along with feature articles and historical narratives, she's published poetry and peer-reviewed academic work. In addition to *Icebound*, she's the author of *One Long Night: A Global History of Concentration Camps* (2017) and *The Secret History of Vladimir Nabokov* (2013).

Andrea has spoken about her writing at the 92nd Street Y and Smithsonian Associates, and delivered panel presentations at the Modern Language Association, the International Journalism Festival, and the Association of Writers & Writing Programs. She's also lectured on history and narrative journalism in the United States and abroad.

Events and ideas that were once common knowledge but have fallen from public memory fascinate Andrea, as does humanity's tendency to not learn from the past. Though she's reported from four continents—from Chile and Myanmar to the Arctic—she feels most at home in libraries or on a boat in the far north.

In 2009, at the Nieman Foundation for Journalism at Harvard, she founded the narrative nonfiction site Nieman Storyboard, which she edited for three years. Before that, she was a freelance journalist, a music critic, a portrait painter, a French translator, a record store manager, and a martial-arts and self-defense instructor (but not all at the same time). She once stopped a runaway bus from crashing, but it wasn't as exciting as it sounds.